SpringerBriefs in Mathematics

SpringerBriefs in Mathematics showcases expositions in all areas of mathematics and applied mathematics. Manuscripts presenting new results or a single new result in a classical field, new field, or an emerging topic, applications, or bridges between new results and already published works, are encouraged. The series is intended for mathematicians and applied mathematicians.

More information about this series at http://www.springer.com/series/10030

Oscar E. Lanford III · Michael Yampolsky

Fixed Point of the Parabolic Renormalization Operator

 Springer

Oscar E. Lanford III
(1940–2013)

Michael Yampolsky
Department of Mathematics
University of Toronto
Toronto, ON
Canada

Oscar E. Lanford III (1940–2013)—Deceased

ISSN 2191-8198   ISSN 2191-8201   (electronic)
ISBN 978-3-319-11706-5   ISBN 978-3-319-11707-2   (eBook)
DOI 10.1007/978-3-319-11707-2

Library of Congress Control Number: 2014951732

Mathematics Subject Classification (2010): 37F25, 37F10

Springer Cham Heidelberg New York Dordrecht London

Printed on acid-free paper

Springer is part of Springer Science+Business Media (www.springer.com)

Preface

This project was born during my sabbatical visit to ETH Zürich in Spring of 2007, and the first version of this manuscript appeared as a preprint in 2011. Sadly, I had to complete it alone; Oscar Lanford passed away in November 2013. Collaborating with Oscar was a true joy, and I miss him both as a colleague and as a friend. In preparing the text for publication, I have attempted not to introduce changes into Oscar's writing style. Without his meticulous attention to detail, I may have overlooked some bugs—hopefully, all of these are minor and will not detract from the mathematical content of the book. I invite the reader to enjoy the depth of Oscar's insight into renormalization and the use of computer-assisted methods in its study. I am grateful for having had the opportunity to work with him, I only wish that we were able to finish this work together.

Toronto, June 2014 Michael Yampolsky

Contents

Chapter 1
Introduction

Abstract The introduction contains a brief and informal description of the concept of parabolic renormalization and of our main results.

Keyword Inou-Shishikura renormalization fixed point

Parabolic renormalization was first introduced by Shishikura in his celebrated work [SH] on the Hausdorff dimension of the boundary of the Mandelbrot set. More recently, the result of Inou and Shishikura [IS] on the convergence of parabolic renormalization became a key to the construction of quadratic Julia sets of positive measure by Buff and Chéritat [BS]. Thus, parabolic renormalization is clearly a powerful and important tool; indeed, it is one of the most important analytic tools to emerge in studying the measure and dimension of Julia sets. Yet it remains one of the more difficult and subtle chapters of modern Complex Dynamics, still imperfectly understood and in many ways mysterious.

Indeed, even the definition of parabolic renormalization is quite complicated. Skipping all of the (important) details, we attempt to summarize it below, as follows. Start with a simple parabolic germ of an analytic function at the origin of the form

$$f(z) = z + a_2 z^2 + \sum_{n \geq 3} a_n z^n, \quad \text{with } a_2 \neq 0.$$

Using a linear change of coordinates we assure without loss of generality that $a_2 = 1$. The classical Leau-Fatou flower theorem (presented with great care in e.g. [Mil1]) describes the local dynamics of f near the origin as follows. There exists a topological disk P_A, known as an attracting petal of f, such that $f(P_A) \subset P_A \cup \{0\}$, and the iterates $f^n(z)$ converge to 0 uniformly for $z \in P_A$. Moreover, every orbit of f which converges to 0 eventually lands in P_A. A repelling petal P_R of f is defined to be an attracting petal for the local branch of f^{-1} which fixes the origin. Together, P_R and P_A form a punctured neighborhood of the origin.

© The Author(s) 2014   1
O.E. Lanford III and M. Yampolsky, *Fixed Point of the Parabolic
Renormalization Operator*, SpringerBriefs in Mathematics,
DOI 10.1007/978-3-319-11707-2_1

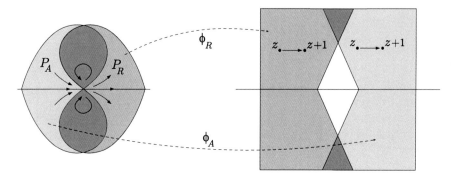

Fig. 1.1 Attracting and repelling petals of a simple parabolic germ of the form $f(z) = z + z^2 + \cdots$, and the corresponding Fatou coordinates

Inside a petal, the dynamics of f can be linearized. There exists a conformal map

$$\phi_A : P_A \to \mathbb{C},$$

which solves the Abel functional equation

$$\phi_A(f(z)) = \phi_A(z) + 1.$$

The map ϕ_A is known as an *attracting Fatou coordinate*; it is defined uniquely up to an addition of a constant. A *repelling Fatou coordinate* ϕ_R is similarly defined as a solution of the Abel equation in a repelling petal. The petals can be chosen so that the image $\phi_A(P_A)$ contains a sector $\{|\mathrm{Arg}(z - C)| < \pi/2 + \varepsilon\}$ for a sufficiently large $C > 0$ and some $\varepsilon > 0$, and similarly the image $\phi_R(P_R)$ contains a sector $\{|\mathrm{Arg}(z + C)| > \pi/2 - \varepsilon\}$ (Fig. 1.1).

By design, the composition $\phi_A \circ (\phi_R)^{-1}$ commutes with the unit translation $z \mapsto z + 1$. Using this fact, it is not difficult to see that it is defined for all z with $|\mathrm{Im} z| > M$ for a sufficiently large value of M. If we denote

$$\mathrm{ixp}(z) \equiv \exp(2\pi i z),$$

then the composition

$$\mathbf{h} \equiv \mathrm{ixp} \circ \phi_A \circ (\phi_R)^{-1} \circ \mathrm{ixp}^{-1} \tag{1.1}$$

defines a pair of analytic maps h^+, h^- defined in punctured neighborhoods of 0 and ∞ respectively. These maps have removable singularities at 0 and ∞, and $h^+(0) = 0$, $h^-(\infty) = \infty$. The multipliers of the fixed points 0 and ∞ are both nonzero.

The pair of analytic germs of h^+ at 0 and h^- at ∞ is Voronin's form of the Écalle-Voronin conformal conjugacy invariant of f [Ec, Vor]. These germs are not quite uniquely defined: a choice of additive constants in the definition of ϕ_A, ϕ_R induces a pre- and post-composition of $\mathbf{h}$ with multiplications by nonzero complex numbers. Now let us say that f is *renormalizable* if

$$h^+(z) = b_1 z + b_2 z^2 + \cdots$$

with $b_2 \neq 0$. Then there is a unique choice of nonzero constants α, β for which

$$\alpha h^+(\beta z) = z + z^2 + \cdots . \tag{1.2}$$

We call the germ (1.2) the *parabolic renormalization* of f.

The term *renormalization* has an established meaning in dynamics: it stands for a rescaled first return map. Although it is not at all obvious from the above description, parabolic renormalization can be interpreted as a limiting case of an appropriately conformally rescaled renormalization of almost parabolic germs (see e.g. [IS]).

In [IS], Inou and Shishikura demonstrated that the successive parabolic renormalizations $\mathscr{P}^n(f_0)$ of the quadratic polynomial $f_0(z) = z + z^2$ converge to an analytic map f_* defined in a neighborhood of the origin, which satisfies the fixed point equation

$$\mathscr{P}(f_*) = f_* . \tag{1.3}$$

They proved that in a suitably restricted class of maps, f_* is a globally attracting fixed point of $\mathscr{P}$.

We note that in general the Eq. (1.3) has many different solutions. Indeed, Schäfke (private communication) has recently described a nondynamical construction of fixed points of the operator $\mathscr{P}$ with an arbitrarily specified h^-. However, if our germ f extends to an analytic map with nice global covering properties, then, generically, its renormalizations will converge to f_*.

This work grew out of our efforts to provide a natural geometric description for the class of maps invariant under $\mathscr{P}$ (and, in particular, for the fixed point f_* itself), and to carry out a computer-assisted study of f_* and $\mathscr{P}$. We describe a natural class of analytic maps $\mathbf{P}_0$ which have a maximal analytic extension to a Jordan domain satisfying the invariance property

$$\mathscr{P} : \mathbf{P}_0 \rightarrow \mathbf{P}_0 .$$

The covering properties of a map $f \in \mathbf{P}_0$ admit an explicit topological model, which we describe in some detail. We prove that the Inou-Shishikura fixed point f_* of $\mathscr{P}$ is contained in $\mathbf{P}_0$, and use the convergence result of [IS] to show that successive renormalizations of any map $f \in \mathbf{P}_0$ converge to f_*.

When it comes to a numerical study of the action of $\mathscr{P}$, one encounters an immediate challenge: estimating the attracting and repelling Fatou coordinates of an analytic germ in the definition (1.1) with sufficient precision. We approach their computation from a new angle, utilizing an asymptotic series for a Fatou coordinate. The existence of such an asymptotic series has at least in some cases been known from the work of Écalle [Ec] on Resurgence Theory. We give an elementary analytic proof of this fact in the general case. We then use the asymptotic series to design a computational scheme for $\mathscr{P}$, and use it to compute with high accuracy Taylor's expansion of f_*, as well as to compute the boundary of its maximal domain of analyticity. We also produce an explicit estimate of the spectral radius of the linearization $D\mathscr{P}|_{f_*}$ in a suitable Banach ball.

We have strived to make our exposition self-contained. In Chap. 2 the reader will find a detailed exposition of the local theory of simple parabolic germs. In addition to standard material, in Sect. 2.2 we present a proof of the existence of an asymptotic series of the Fatou coordinate at infinity, which plays a key role in our numerical experiments. We make a brief note of the role this series plays in Écalle's Resurgence Theory for Fatou coordinates in Sect. 2.2.1. In Sect. 2.3, after a discussion of Écalle-Voronin conformal conjugacy invariants, the parabolic renormalization operator $\mathscr{P}$ makes its first appearance. Chapter 3 discusses global properties of parabolic renormalization, starting with a detailed discussion of parabolic renormalization of the quadratic polynomial $f_0(z) = z + z^2$ in Sect. 3.2. We define the class $\mathbf{P_0}$ in Sect. 3.5. Section 3.6 contains the core of the proof of invariance of $\mathbf{P_0}$ under the action of $\mathscr{P}$. In Sect. 3.7 we use the results of Inou and Shishikura to show that parabolic renormalizations of a map in $\mathbf{P_0}$ converge to the Inou-Shishikura fixed point f_*. Chapter 4 contains a numerical study of the parabolic renormalization operator and its fixed point f_*. Here we make use of the asymptotic series for the Fatou coordinates of a parabolic germ. Some amusing examples are left for dessert in Chap. 5.

Chapter 2
Local Dynamics of a Parabolic Germ

Abstract We discuss the local dynamics of a simple parabolic germ and Fatou coordinates, the asymptotics of a Fatou coordinate at infinity, and the definition and basic properties of the parabolic renormalization of a simple parabolic germ.

Keywords Simple parabolic germ · Fatou coordinate · Resurgent series · Écalle-Voronin invariants · Parabolic renormalization

2.1 Fatou Coordinates

We briefly review the local dynamics of an analytic function f in the vicinity of a parabolic fixed point at 0:

$$f(z) = e^{2\pi i p/q} z + O(z^2).$$

We consider first the case $q = 1$, that is, $f'(0) = 1$, and we write

$$f(z) = z + az^{n+1} + O(z^{n+2}), \tag{2.1}$$

for some $n \in \mathbb{N}$ and $a \neq 0$. The integer $n + 1$ can be recognized as the *multiplicity* of 0 as the solution of $f(z) - z = 0$.

A complex number v of modulus one is called an *attracting direction* if the product $av^n < 0$, and a *repelling direction* if the same product is positive. The terminology has the following meaning.

Proposition 2.1 *Let $\{f^k(z)\}$ be an orbit in $\mathrm{Dom}(f) \setminus \{0\}$ which converges to the parabolic fixed point 0. Then the sequence of unit vectors $f^k(z)/|f^k(z)|$ converges as $k \to \infty$ to one of the attracting directions.*

We say in this case that the orbit converges to p *from the direction of v*.

If f has a parabolic fixed point at 0, it admits a *local inverse* there, by which we mean a function g, defined and analytic in a neighborhood of 0, so that

© The Author(s) 2014
O.E. Lanford III and M. Yampolsky, *Fixed Point of the Parabolic
Renormalization Operator*, SpringerBriefs in Mathematics,
DOI 10.1007/978-3-319-11707-2_2

$g(f(z)) = z = f(g(z))$ for z near enough to 0. The germ at 0 of a local inverse is unique, but its domain of definition typically has to be chosen. A local inverse also has a parabolic fixed point at 0; attracting directions for f are repelling for the inverse and vice versa.

Definition 2.1 Let v be an attracting direction for f. An *attracting petal for f (from the direction v)* is a Jordan domain P with closure in Dom(f) such that:

1. $0 \in \partial P$;
2. f is injective on P;
3. $f(\overline{P} \setminus \{0\}) \subset P$;
4. for any $z \in P$, the orbit $f^k(z)$ converges to 0 from the direction v, and the convergence of f^k to 0 is uniform on P;
5. conversely, any orbit $f^k(z)$ which converges to z from the direction v is eventually in P.

Similarly, U is a *repelling petal* for f if it is an attracting petal for some local inverse g of f.

Judiciously chosen petals can be organized into a *Leau-Fatou Flower* at 0:

Theorem 2.1 *There exists a collection of n attracting petals P_i^a, and n repelling petals P_j^r such that the following holds. Any two repelling petals do not intersect, and every repelling petal intersects exactly two attracting petals. Similar properties hold for attracting petals. The union*

$$(\cup P_i^a) \cup (\cup P_j^r) \cup \{0\}$$

forms an open simply-connected neighborhood of 0.

The proof of this statement relies on some changes of coordinates. First: every germ of the form (2.1) can be brought into the form

$$f(z) = z + z^{n+1} + \alpha z^{2n+1} + O(z^{3n+1}) \tag{2.2}$$

in a suitable conformal local coordinate change at 0. In fact, a straightforward induction shows the following:

Proposition 2.2 (cf. [Mil1] Problem 10-d, [BE]) *For every germ of the form (2.1) there exists a unique $\alpha \in \mathbb{C}$ such that for every $N \in \mathbb{N}$ greater than $2n + 1$ there is a locally conformal change of coordinates ψ, with $\psi(0) = 0$, such that*

$$\psi \circ f \circ \psi^{-1}(z) = z + z^{n+1} + \alpha z^{2n+1} + O(z^N).$$

Further, there exists a formal power series $\Psi(z) = \sum_{k=1}^{\infty} p_k z^k$ that formally conjugates

$$\Psi \circ f \circ \Psi^{-1}(z) = z + z^{n+1} + \alpha z^{2n+1}.$$

Thus, the number $\alpha \in \mathbb{C}$ is a formal conjugacy invariant of f, and specifies its formal conjugacy class uniquely.

For the next few paragraphs, we will take f to have the special form (2.2). The attracting directions are then the nth roots of -1. We will describe some ways of construction attracting petals for the attracting direction v; the adjustments necessary to deal with repelling petals are routine. The reader is reminded that (2.2) is *not* the general form of a mapping with a parabolic fixed point of order n; it has been cleaned up by making a preliminary analytic change of coordinates to eliminate some powers of z in its Taylor series.

The behavior of orbits of such an f near 0 is greatly clarified by making the coordinate change

$$w = \kappa(z) := -\frac{1}{nz^n}.$$

We are considering a particular attracting direction, and we take κ to be defined on the sector between the two adjacent repelling directions; it opens up this sector to the complex plane cut along the positive real axis. With its domain of definition restricted in this way, κ is bijective, and its inverse is given by

$$\kappa^{-1}(w) = \left(-\frac{1}{nw}\right)^{1/n},$$

where the branch of the nth root is the one cut along the positive real axis and taking the value v at -1.

Then

$$F(w) := \kappa \circ f \circ \kappa^{-1}(w)$$

$$= -\frac{1}{n}\left(f\left(\left(-\frac{1}{nw}\right)^{\frac{1}{n}}\right)\right)^{-n}$$

$$= -\frac{1}{n}\left(\left(-\frac{1}{nw}\right)^{\frac{1}{n}} + \left(-\frac{1}{nw}\right)^{\frac{n+1}{n}} + \alpha\left(-\frac{1}{nw}\right)^{\frac{2n+1}{n}} + O\left(\left(\frac{-1}{nw}\right)^{\frac{3n+1}{n}}\right)\right)^{-n}$$

$$= w \cdot \left(1 + \left(-\frac{1}{nw}\right) + \alpha\left(\frac{1}{nw}\right)^2 + O\left(\frac{1}{w^3}\right)\right)^{-n}$$

$$= w \cdot \left(1 + \frac{1}{w} - \frac{\alpha}{n} \cdot \frac{1}{w^2} + \frac{n+1}{2n} \cdot \frac{1}{w^2} + O(\frac{1}{w^3})\right).$$

We thus obtain

$$F(w) = w + 1 + \frac{A}{w} + O(1/w^2) \quad \text{as } w \to \infty$$

where

$$A = \frac{1}{n}\left(\frac{n+1}{2} - \alpha\right).$$

Selecting a right half-plane $H_r = \{\text{Re}\, z > r\}$ for a sufficiently large $r > 0$, we have

$$\text{Re}\, F(w) > \text{Re}\, w + 1/2, \quad \text{and hence } F(\overline{H_r}) \subset H_r.$$

The domain $\kappa^{-1}(H_r)$ is then an attracting petal for the attracting direction v. In the case of a simple parabolic point, what we have just shown simplifies to the assertion that any disk of sufficiently small radius, tangent to the imaginary axis from the left at 0, is an attracting petal.

The petals just discussed—pullbacks of half-planes under κ—have boundaries tangent at the origin to directions $e^{\pm i\pi/(2n)}v$. For many purposes, such as the proof of Theorem 2.1, we will need petals with a strictly larger opening angle. There are many ways to construct such petals; here is one which is convenient for our purposes. Let $\pi/2 < \alpha < \pi$, $R > 0$, and let

$$\Delta(\alpha, R) := \{w : -\alpha < \text{Arg}(w - R) < \alpha\} \tag{2.3}$$

(i.e., $\Delta(\alpha, R)$ is the sector $\{-\alpha < \text{Arg}(w) < \alpha\}$ translated right by R.) From

$$F(w) = w + 1 + O(1/w),$$

there exists a $R_0 = R_0(\alpha)$ so that

$$|\text{Arg}(F(w) - w)| < \pi - \alpha \quad \text{and} \quad \text{Re}(F(w)) > \text{Re}(w) + 1/2 \tag{2.4}$$

for $|w| \geq R_0$. If R is large enough, the domain $\Delta(\alpha, R)$ does not intersect the disk $\{|w| \leq R_0\}$, so (2.4) holds for $w \in \Delta(\alpha, R)$. For such R's, by elementary geometric considerations,

$$F\left(\overline{\Delta(\alpha, R)}\right) \subset \Delta(\alpha, R),$$

and $F^n(w) \xrightarrow[n\to\infty]{} \infty$ for all $w \in \Delta(\alpha, R)$. Further any $\Delta(\alpha, R)$ contains a right half-plane and hence eventually contains any F-orbit converging to ∞. Finally, it can be verified that the sequence of iterates F^n converges uniformly to ∞ on $\Delta(\alpha, R)$. We omit this verification; it uses simplified versions of the ideas used in the proof of Lemma 2.1. Thus, sets of the form $\kappa^{-1}(\Delta(\alpha, R))$ are attracting petals, symmetric about the attracting direction under consideration, with tangents at the origin in directions $e^{\pm i\alpha/n}v$. It will be useful to have a general term for behavior for this: We will say that a petal P with attracting or repelling direction v is *ample* if

$$P \supset \{z : |\mathrm{Arg}(z/v)| < \alpha/n, |z| < r\}$$

for some $\alpha > \pi/2$ and sufficiently small r.

The dynamics inside a petal is described by the following:

Proposition 2.3 *Let P be an attracting petal for f. Then there exists a conformal change of coordinates ϕ defined on P, conjugating $f(z)$ to the unit translation $T : z \mapsto z + 1$.*

Proof For a traditional proof, see e.g., [Mil1] §10. We cannot resist giving a proof based on quasiconformal surgery, which probably originated in the work of Voronin [Vor]. For definiteness, we discuss the case of an attracting petal with attracting direction v, and let

$$F(w) = w + 1 + O(1/w)$$

be as above. Also as above, we select a right half-plane H_r. The main step will be to prove the existence of ϕ for the special petal $\kappa^{-1} H_r$, which we provisionally denote by P_0. The case of a general petal will then follow by an easy extension argument.

As we know, $F(\overline{H_r}) \subset H_r$, so let us denote S to be the closed strip

$$S = \overline{H_r \setminus F(H_r)}.$$

Setting $\mathbb{S} = \{\mathrm{Re}\, z \in [0, 1]\}$, let h be any diffeomorphism

$$h : S \to \mathbb{S},$$

which on the boundary of the strip conjugates F to T:

$$T \circ h(w) = h \circ F(w), \quad \text{for all } w \text{ with } \mathrm{Re}\, w = r.$$

We will further require that the first partial derivatives of h and h^{-1} be uniformly bounded in S. Verifying the existence of a diffeomorphism with these properties is an elementary exercise which we leave to the reader.

The diffeomorphism h defines a new complex structure $\mu = h_* \sigma_0$ on $\mathbb{S}$, which we extend to the left half-plane $\{\mathrm{Re}\, z < 0\}$ by

$$\mu|_w = (T^n)^* \mu \text{ for } T^n(w) \in \mathbb{S}.$$

Gluing together H_r with the standard complex structure and the half-plane $\{\mathrm{Re}\, z < 1\}$ with structure μ via the homeomorphism h (which is now analytic), and using the Measurable Riemann Mapping Theorem, we obtain a new Riemann surface X. By the Uniformization Theorem, X is conformally isomorphic either to $\mathbb{C}$ or to the disk. By construction, X is quasiconformally isomorphic to $\mathbb{C}$ and therefore cannot be conformally isomorphic to the disk. We can specify a conformal isomorphism $\Phi : X \to \mathbb{C}$ uniquely by imposing normalization conditions $\Phi(0) = 0$ and $\Phi(-1) = -1$.

The pair of maps $T|_{\{\mathrm{Re}\, z<0\}}$ and $F|_{H_r}$ induces a conformal automorphism of X, which we denote by $\tilde{F}$. Then $\Phi \circ \tilde{F} \circ \Phi^{-1}$ is a conformal automorphism of $\mathbb{C}$ with no fixed point. It is a standard fact that the only such automorphisms are translations, and our choice of normalization for Φ implies that

$$\Phi \circ \tilde{F} \circ \Phi^{-1} \equiv T. \tag{2.5}$$

But $\tilde{F} = F$ on $H_r \subset X$, so we get

$$\Phi \circ F = T \circ \Phi \quad \text{on } H_r.$$

Moreover, the restriction of Φ to H_r is analytic in the standard sense. Thus, we set $\phi := \Phi \circ \kappa$ on $P_0 := \kappa^{-1} H_r$ and obtain

$$\phi \circ f = \phi + 1 \quad \text{on } P_0,$$

as desired. Since Φ is a conformal isomorphism from X to $\mathbb{C}$, the map ϕ is univalent on P_0.

This proves the existence of ϕ on the particular petal P_0. We provisionally denote the above ϕ, which is defined on P_0, by ϕ_0. We define

$$B_v^f := \{z : f^n(z) \to 0 \quad \text{from the direction } v\}.$$

If $z_0 \in B_v^f$, then $f^n(z_0) \in P_0$ for sufficiently large n. If $f^{n_0}(z_0) \in P_0$, then $(f^{n_0})^{-1} P_0$ is an open set containing z_0 and contained in B_v^f, so B_v^f is open. Since P_0 is mapped into itself by f, and since

$$\phi_0(f(z)) = \phi_0(z) + 1 \quad \text{on } P_0,$$

$\phi_0(f^n(z_0)) - n$ takes the same value for all n for which $f^n(z_0) \in P_0$. We denote this common value by $\phi(z_0)$, thus obtaining a function ϕ defined on all of B_v^f and extending ϕ_0 defined on $P_0 \subset B_v^f$. Tautologically,

$$\phi(f(z)) = \phi(z) + 1.$$

If $f^{n_0}(z_0) \in P_0$, then $f^{n_0}(z) \in P_0$ on a neighborhood of z_0, so $\phi(z) = \phi_0(f^{n_0}(z)) - n_0$ on this neighborhood, which shows that the extended ϕ is analytic, but not necessarily univalent, on all of B_v^f.

Now let P be a general attracting petal with the same attracting direction v. By the definition of a petal, $P \subset B_v^f$, so we can restrict ϕ to P, thus obtaining an analytic function satisfying

$$\phi(f(z)) = \phi(z) + 1 \quad \text{on } P.$$

It remains to show that the restriction of ϕ to P is univalent. To see this, let z_1, z_2 be points of P with $\phi(z_1) = \phi(z_2)$. For sufficiently large n, $f^n(z_1)$ and $f^n(z_2)$ are both in P_0, so

$$\phi_0(f^n(z_1)) = \phi(z_1) + n = \phi(z_2) + n = \phi_0(f^n(z_2)).$$

But, by construction, ϕ_0 is univalent, so $f^n(z_1) = f^n(z_2)$. The argument so far works for any pair z_1, z_2 in B_v^f with $\phi(z_1) = \phi(z_2)$. Now, however, we use that facts that z_1 and z_2 are both in the petal P, which is mapped into itself by f and on which f is univalent. Hence, from $f^n(z_1) = f^n(z_2)$ it follows that $z_1 = z_2$, proving univalence of ϕ on P. □

We note for future reference a simple result which was proved in the course of the preceding argument.

Proposition 2.4 *Let v be an attracting direction for f, let P be an attracting petal from the direction v, and let ϕ a univalent analytic function defined on P and satisfying the function equation*

$$\phi(f(z)) = \phi(z) + 1.$$

Then ϕ has a unique extension to B_v^f satisfying this equation.

We define *attracting Fatou coordinate* (for the attracting petal P with attracting direction v) to be a function $\phi_A : P \to \mathbb{C}$ analytic and univalent on P and satisfying

$$\phi_A(f(z)) = \phi_A(z) + 1 \quad \text{on } P.$$

As we have seen, such a function extends uniquely, via the above functional equation, to all of B_v^f, and the extension restricts to an attracting Fatou coordinate on any other petal with attracting direction v. It is clear that if ϕ_A is an attracting Fatou coordinate, then so is $\phi_A + c$ for any constant c. We will see shortly that any two attracting Fatou coordinates differ only in this way.

Any attracting Fatou coordinate can be written in the form $\phi_A = \Phi_A \circ \kappa$, where Φ_A satisfies the functional equation

$$\Phi_A(F(w)) = \Phi_A(w) + 1 \quad \text{(with } F = \kappa \circ f \circ \kappa^{-1} \text{ as above)}$$

on an appropriate F-invariant domain "near infinity". We will refer to such Φ_A's as *Fatou coordinates at infinity*.

A *repelling Fatou coordinate* ϕ_R for f means an attracting Fatou coordinate for an analytic local inverse g of f. If

$$f(z) = z + z^{n+1} + \cdots, \text{ then } g(z) = z - z^{n+1} + \cdots,$$

which can be brought back into the standard form by conjugating with $z \mapsto -z$ (n odd) or $z \mapsto i \cdot z$ (n even). The above considerations then apply to define ϕ_R, repelling petals, etc. We note that:

- a repelling Fatou coordinate ϕ_R satisfies *the same* functional equation

$$\phi_R(f(z)) = \phi_R(z) + 1$$

  as does an attracting one, but the domains of definition are different, and
- the image of a repelling petal by a repelling Fatou coordinate is mapped into itself by the unit *left* translation $w \mapsto w - 1$; the image of an ample repelling petal under a repelling Fatou coordinate contains a *left* half-plane.

Again, it is useful to also consider *repelling Fatou coordinates at infinity*: If ϕ_R is a repelling Fatou coordinate, the corresponding one at infinity is

$$\Phi_R(w) = \phi_R(\kappa^{-1}(w))$$

(but the appropriate branches of κ^{-1} are different from the ones in the attracting case).

Our next step is to prove a crude asymptotic formula for a Fatou coordinate at infinity. It is advantageous here to deviate from what we have been doing. We consider a mapping f of the form

$$f(z) = z + z^{n+1} + f_{n+2}z^{n+2} + \cdots,$$

i.e., we do not assume we have made a preliminary change of variable to eliminate, e.g., the terms z^j for j between $n + 1$ and $2n + 1$. We introduce $F = \kappa \circ f \circ \kappa^{-1}$ as before; this time, the behavior of F near infinity is

$$F(w) = w + v(w) \quad \text{where } v(w) = 1 + v_1 w^{-1/n} + v_2 w^{-2/n} + \cdots;$$

the series converges for sufficiently large $|w|$. Let Φ_A be an attracting Fatou coordinate at infinity. By what we have already proved, for any $\alpha < \pi$, there is an R so that Φ_A extends analytically to a univalent function on the set $\{|w| > R, -\alpha < \text{Arg}(w) < \alpha\}$.

Proposition 2.5
$$\Phi_A'(w) \to 1 \quad and \quad \Phi_A(w)/w \to 1$$

uniformly as $w \to \infty$ in any sector $\{-\alpha < \text{Arg}(w) < \alpha\}$ with $\alpha < \pi$. The same limits hold for Φ_R, but with $w \to \infty$ in the opposite sector $\{-\alpha < \text{Arg}(-w) < \alpha\}$

This proposition is a less-precise version of Lemma A.2.4 of [Sh], and the argument we give is the first part of Shishikura's proof of that lemma. Shishikura carries the analysis further and is able to identify, in favorable cases, the first correction to the indicated asymptotic behaviors. We do not give his full argument here, as we will prove Theorem 2.2, which gives more precise information about the asymptotic behavior of Fatou coordinates.

Proof Fix α with $\pi/2 < \alpha < \pi$, and let $\alpha < \alpha_1 < \pi$. Take R_1 so that Φ_A is defined and univalent in

$$S_1 := \{|w| > R_1, -\alpha_1 < \text{Arg}(w) < \alpha_1\}$$

and also so that $|v(w) - 1| < 1/4$ on S_1. For $w_0 \in S_1$, denote by $\rho = \rho(w_0)$ the distance from w_0 to the boundary of S_1. We will investigate limits as $|w_0| \to \infty$ in the strictly smaller sector $-\alpha < \text{Arg}(w_0) < \alpha$; then there is a constant $k > 0$ so that, asymptotically, $\rho(w_0) \geq k \cdot |w_0|$. In the following, we will frequently assume that $\rho(w_0)$ is "large enough". We will also use C to denote a generic "universal" constant; different instances of C do not have to denote the same constant.

For the first step, we use the Koebe Distortion Theorem. If $|w - w_0| < \rho - 2$, so the disk of radius 2 about w is in S_1, then the mapping

$$a \mapsto \frac{\Phi_A(w + a) - \Phi_A(w)}{\Phi'_A(w)}$$

is analytic and univalent on $\{|a| < 2\}$ and has unit derivative at the origin. A simple rescaling of the Koebe Theorem to adapt it to the disk of radius 2 gives a universal constant $C > 0$ so that

$$\left| \frac{\Phi_A(w + a) - \Phi_A(w)}{\Phi'_A(w)} \right| > C^{-1} \quad \text{for } 3/4 < |a| < 5/4.$$

We insert $a = v(w)$ into this estimate, use $|1 - v(w)| < 1/4$ to ensure that $3/4 < |v(w)| < 5/4$, and use also the functional equation

$$\Phi_A(w + v(w)) = \Phi_A(F(w)) = \Phi_A(w) + 1$$

to get

$$|\Phi'_A(w)| < C \quad \text{for } |w - w_0| < (\rho - 2).$$

Applying the Cauchy estimates gives a bound

$$|\Phi''_A(w)| \leq \frac{C}{\rho} \quad \text{for } |w - w_0| < \rho/2$$

(with a different C).

Next we apply Taylor's Formula with Integral Remainder to write

$$\Phi_A(w_0 + a) = \Phi_A(w_0) + a \cdot \Phi'(w_0) + a^2 \cdot \int_{s=0}^{1} (1 - s)\Phi''(w_0 + s \cdot a)ds.$$

Again, we set $a = v(w_0)$ and use $\Phi_A(w_0 + v(w_0)) = \Phi_A(w_0) + 1$ to get

$$
1 - v(w_0)\Phi_A'(w_0) = (v(w_0))^2 \int\limits_{s=0}^{1} (1 - s)\Phi''(w_0 + s \cdot a)ds.
$$

Since the estimate $|\Phi_A''(w)| \le C/\rho$ holds for all w appearing in the integral on the right, we get

$$
|1 - v(w_0) \cdot \Phi_A'(w_0)| \le C/\rho(w_0).
$$

We have already remarked that $\rho(w_0) \ge k|w_0|$ as $w_0 \to \infty$ in the sector

$$
\{-\alpha < \mathrm{Arg}(w_0) < \alpha\},
$$

so

$$
|\Phi_A'(w_0) - v(w_0)^{-1}| = O(|w_0|^{-1}) \quad \text{in that sector.}
$$

This establishes the asserted convergence of Φ_A'; the assertion about $\Phi(w)/w$ follows by integration. $\qquad\square$

Equipped with this information about the asymptotic behavior of Fatou coordinates, we can now show that the image of a Fatou coordinate is large enough. As usual, it suffices, up to insertion of some minus signs, to consider the attracting case. Let Φ_A be an attracting Fatou coordinate, and let $0 < \alpha < \pi$. Then, for sufficiently large R, Φ_A extends analytically to a univalent function on

$$
S := \{w : |w| > R, -\alpha < \mathrm{Arg}(w) < \alpha\}.
$$

Proposition 2.6 *Let $0 < \alpha_0 < \alpha$. Then, for sufficiently large R_0,*

$$
\Phi_A(S) \supset S_0 := \{|w| < R_0, -\alpha_0 < \mathrm{Arg}(w) < \alpha_0\}.
$$

Proof Let $w_0 \in S_0$; we want to investigate solutions to the equation $\Phi_A(w) = w_0$ which we rewrite as

$$
w = w_0 + w - \Phi_A(w) =: \Psi_0(w).
$$

The idea is to apply the Contraction Mapping Principle to Ψ_0, using the fact that $\Psi_0'(w) = 1 - \Phi_A'(w)$ which is small for w large. To do this, we need to find a domain mapped into itself by Ψ_0 and on which Ψ_0 is contractive. Suppose we can find a $\delta > 0$ so that

- $|\Psi_0'(w)| < 1/2$ for $|w - w_0| < \delta$
- $|\Psi_0(w_0) - w_0| < \delta/2$.

Then, for $|w - w_0| < \delta$,

$$|\Psi_0(w) - w_0| \leq |\Psi_0(w) - \Psi_0(w_0)| + |\Psi_0(w_0) - w_0|$$
$$< (1/2) \cdot |w - w_0| + \delta/2 \leq \delta/2 + \delta/2 = \delta,$$

so the disk of radius δ about w_0 will be mapped to itself, and Ψ_0 will have a unique fixed point in this disk.

We implement this strategy as follows. First, we arrange, by making R larger if necessary, that $|\Phi'_A(w) - 1| < 1/2$ on S. We write

$$\varepsilon := \sin(\alpha - \alpha_0),$$

and we note that, by elementary geometry, $|w - w_0| < \varepsilon \cdot w_0$ implies

$$|\mathrm{Arg}(w/w_0)| < \alpha - \alpha_0.$$

If we further take $R_0 \geq (1 - \varepsilon)^{-1} R$, then the disk of radius $\delta := \varepsilon \cdot |w_0|$ about w_0 is contained in S, for any $w_0 \in S_0$. Recall that we have already arranged that $|\Psi'_0| < 1/2$ on S. Finally, we apply Proposition 2.5 to see that by taking R_0 large enough, we can arrange that

$$|\Phi_A(w_0) - w_0| < (1/2) \cdot \varepsilon \cdot |w_0| \quad \text{for all } w_0 \in S_0.$$

All the elements for the above contraction argument are now in place, and we can conclude that for every $w_0 \in S_0$, there is a unique w with $\Phi_A(w) = w_0$ in

$$\{|w - w_0| < \varepsilon|w_0|\} \subset S.$$

This proves the assertion $\Phi_A(S) \supset S_0$.   □

It follows from this proposition that we have

Proposition 2.7 *The image under ϕ_A of any ample petal of f contains a right half-plane.*

A similar assumption holds for ample repelling petals, but with the image under ϕ_R covering a left half-plane.

Let P be an attracting petal. We define a relation on P by $z_1 \sim z_2$ if z_1 and z_2 are on the same orbit, that is, if either $z_2 = f^j(z_1)$ or $z_1 = f^j(z_2)$ (with $j \geq 0$). It is easy to check that this is an equivalence relation.

Consider the quotient $P/\sim$. The canonical projection $\pi : P \to P/\sim$ is locally injective, and it is straightforward to verify that there is a unique way to give $P/\sim$ a Riemann surface structure, in such a way as to make π analytic and therefore a local conformal isomorphism.

Let P_1 be another attracting petal contained in P. Since the orbit of every point $z \in P$ eventually lands in P_1, the inclusion $P_1 \hookrightarrow P$ induces a conformal homeomorphism

$$P_1/\sim \xrightarrow{\simeq} P/\sim.$$

Now if P_2 is any attracting petal with the same attracting direction as P, the intersection $P \cap P_2$ is also a petal. Hence

$$P_2/\sim \simeq P/\sim \simeq (P \cap P_2)/\sim.$$

Thus, the quotient $P_\sim$ does not depend on the choice of the petal P but only on the choice of the attracting direction v corresponding to P. We will write

$$P/\sim \equiv \mathscr{C}_A^v.$$

We will omit v from the notation when the choice of the attracting direction is clear from the context (for instance, when there is only one attracting direction).

Let ϕ_A be an attracting Fatou coordinate defined on some petal P. It is easy to verify, using the injectivity of ϕ_A on P, that two points z_1 and z_2 are equivalent if and only if $\phi_A(z_1) - \phi_A(z_2) \in \mathbb{Z}$. Hence, ϕ_A defines, by passage to quotients, an injective mapping $\tilde{\phi}_A$ from $\mathscr{C}_A^v$ to $\mathbb{C}/\mathbb{Z}$. If we take P to be an ample petal, then it follows from Proposition 2.7 that $\tilde{\phi}_A$ takes on all values in $\mathbb{C}/\mathbb{Z}$. Thus:

Proposition 2.8 *The map $\tilde{\phi}_A$ is a conformal isomorphism from the Riemann surface $\mathscr{C}_A^v$ to $\mathbb{C}/\mathbb{Z}$.*

In light of the preceding proposition, we will call $\mathscr{C}_A^v$ the *attracting cylinder* corresponding to the direction v.

The *repelling cylinder* $\mathscr{C}_R^v$ for f is the attracting cylinder for a local inverse of f, fixing the origin.

If P is an attracting petal, we will call the half-open domain

$$C_A = P \setminus fP$$

a *fundamental attracting crescent*, the name reflecting its shape. A fundamental repelling crescent means a fundamental attracting crescent for a local inverse of f.

Proposition 2.9 *For any attracting petal P in the direction v, the fundamental crescent C_A projects bijectively onto the attracting cylinder $\mathscr{C}_A^v$. More concretely: Every point of P lies on a forward orbit starting in C_A, and distinct points of C_A have disjoint forward orbits.*

Proof From the requirement that f^n converge uniformly to 0 on P—condition (4) in of our definition of petal—it follows that no point of P admits an arbitrarily long backward orbit in P. Thus, every point of P lies on the forward orbit starting outside P; the first point on this orbit inside P is in C_A. Let z_1, z_2 be points of C_A whose forward orbits intersect. From the injectivity of f on P, we have, possibly after interchanging z_1 and z_2, that $f^n(z_1) = z_2$ for same $n \geq 0$. If $n > 0$, then $z_2 = f^n(z_1) \in f(P)$, contradicting $z_2 \in C_A$, so the only possibility left is $n = 0$, i.e., $z_1 = z_2$.   □

We note the following standard fact:

Proposition 2.10 *Let $h : \mathbb{C}^* \to \mathbb{C}^*$ be an injective holomorphic map. Then either $h(z) = cz$ or $h(z) = c/z$ for a nonzero constant c.*

Proof Such an h is in particular an analytic function with isolated singularities at 0 and ∞ (and nowhere else). By injectivity, neither singularity can be essential, so h extends to a meromorphic mapping of the Riemann sphere to itself, i.e., to a rational function. Injectivity on the sphere with two points deleted implies that this rational function has degree one, i.e., is a Möbius transformation. In particular, the extended function maps the sphere bijectively to itself, so either $h(0) = 0$, in which case $h(\infty) = \infty$, or $h(0) = \infty$. In the first case, $h(z)/z$ is bounded at ∞ and has a removable singularity at 0, so by Liouville's Theorem, $h(z)/z = c$. In the case $h(0) = \infty$, applying the above to $z \mapsto h(1/z)$ gives $h(z) = c/z$.   □

A corollary of the above result is a uniqueness statement for Fatou coordinates:

Proposition 2.11 *Let P be an attracting petal of f, and let ϕ_1 and ϕ_2 be attracting Fatou coordinates on P, i.e., univalent analytic functions satisfying $\phi_i(f(z)) = \phi_i(z) + 1$. Then $\phi_2(z) - \phi_1(z)$ is constant on P.*

Proof By Proposition 2.8, ϕ_1 and ϕ_2 both induce conformal isomorphisms from $\mathscr{C}_A$ to $\mathbb{C}/\mathbb{Z}$. It will be more convenient to work with the punctured plane $\mathbb{C}^*$ instead of $\mathbb{C}/\mathbb{Z}$. The function

$$\mathrm{ixp}(z) := \exp(2\pi i z)$$

induces, again by passage to quotients, a conformal isomorphism from $\mathbb{C}/\mathbb{Z}$ to C^*, so $\mathrm{ixp} \circ \phi_1$ and $\mathrm{ixp} \circ \phi_2$ both induce conformal isomorphism $\mathscr{C}_A \to \mathbb{C}^*$. Hence, the prescription

$$h : \exp(2\pi i \phi_1(z))) \mapsto \exp(2\pi i \phi_2(z)) \quad \text{for all } z \in P$$

defines a conformal isomorphism $h : \mathbb{C}^* \to \mathbb{C}^*$. By Proposition 2.10, there are two possibilities:

1. there is a nonzero constant, which we write as $\exp(2\pi i c)$, so that

$$\exp(2\pi i c) \cdot \exp(2\pi i \phi_1(z)) = \exp(2\pi i \phi_2(z)) \quad \text{for all } z \in P$$

or

2. there is a constant c so that

$$\exp(2\pi ic) \cdot \exp(-2\pi i\phi_1(z)) = \exp(2\pi i\phi_2(z)) \quad \text{for all } z \in P.$$

In the first case,

$$\phi_2(z) - \phi_1(z) - c \in \mathbb{Z} \quad \text{for all } z \in P.$$

But the expression on the left is continuous, and an integer-valued continuous function on a connected set must be constant, so

$$\phi_2(z) = \phi_1(z) + c + n_0 \quad \text{for all } z \in P, \text{ for some } n_0 \in \mathbb{Z}$$

which is what we wanted to prove.

In the second case, similarly,

$$\phi_2(z) = -\phi_1(z) + c + n_0 \quad \text{for all } z \in P,$$

and this contradicts

$$\phi_1(f(z)) - \phi_1(z) = 1 = \phi_2(f(z)) - \phi_2(z) \quad \text{for } z \in P,$$

so this case is excluded.   □

The critical values of Fatou coordinates are simply related to those of f.

Proposition 2.12 *Let ϕ_A be an attracting Fatou coordinate defined on B_v^f. Then critical values of ϕ_A all have the form*

$$\phi_A(v) - n \quad \text{with } v \text{ a critical value of } f \text{ in } B_v^f \text{ and } n \geq 1.$$

If $f : B_v^f \to B_v^f$ is surjective, all numbers of this form are critical values.

Proof Iterating the functional equation for ϕ_A, differentiating, and applying the chain rule gives

$$\phi_A'(z) = \frac{d}{dz}\phi_A(f^n(z)) = \phi_A'(f^n(z))\prod_{j=0}^{n-1}f'(f^j(z)).$$

For given z and sufficiently large n, $f^n(z)$ is in an attracting petal, which implies $\phi_A'(f^n(z)) \neq 0$. Thus z is a critical point of ϕ_A if and only if there exists $j \geq 0$ such that $f^j(z)$ is a critical point of f. For such a j, $f^{j+1}(z) = v$ is a critical value of f. Since

$$\phi_A(z) = \phi_A(f^{j+1}(z)) - (j+1) = \phi_A(v) - (j+1),$$

the assertion follows.   □

For completeness, let us note how the situation changes if $f'(0)$ is a qth root of unity $e^{2\pi i p/q}$ with $q \neq 1$. A fixed petal for the iterate f^q corresponds to a cycle of q petals for f. It thus follows that q divides the number n of attracting/repelling directions of 0 as a fixed point of f^q.

2.2 Asymptotic Expansion of a Fatou Coordinate at Infinity

We will now specialize to the case $q = 1$ and $n = 1$. By rescaling, we can then bring the coefficient of z^2 to 1, and the normal form (2.2) becomes

$$f(z) = z + z^2 + \alpha z^3 + O(z^4). \tag{2.6}$$

We will say in this case that 0 is a *simple* parabolic fixed point of f. There is one attracting direction (-1) and one repelling direction $(+1)$.

If the domain of definition $\mathrm{Dom}(f) \ni 0$ is fixed, we let $B^f \subset \mathrm{Dom}(f)$, as before, denote the basin of the parabolic point at the origin. The *immediate basin* of 0, which we denote B_0^f, is the connected component of B^f that contains an attracting petal.

The change of variables κ moving the parabolic point to ∞ becomes simply

$$\kappa(z) = -\frac{1}{z}, \quad \kappa^{-1}(w) = -\frac{1}{w},$$

and we have

$$F(w) = -\left(f(-\frac{1}{w})\right)^{-1} = w + 1 + \frac{A}{w} + O(w^{-2}), \quad \text{with } A = 1 - \alpha,$$

and $F(w) - w$ is analytic at ∞.

We showed earlier (Proposition 2.5) that any attracting Fatou coordinate at infinity Φ_A for such an f satisfies

$$\Phi_A(w) = w + o(w) \quad \text{as } w \to \infty \text{ appropriately.}$$

We will prove shortly a much more precise result: an asymptotic expansion giving Φ_A up to corrections of order w^{-n} for any n. Before we do this, we investigate formal solutions to the functional equation

$$\Phi(F(w)) = \Phi(w) + 1,$$

satisfied by both attracting and repelling Fatou coordinates.

Proposition 2.13 *There is a unique sequence $b_1, b_2, \ldots$ of complex coefficients such that*

$$\Phi_{ps}(w) = w - A \log(w) + \sum_{j=1}^{\infty} b_j w^{-j} \tag{2.7}$$

satisfies

$$\Phi_{ps} \circ F(w) = \Phi_{ps}(w) + 1 \quad \text{in the sense of formal power series.} \tag{2.8}$$

Furthermore, if we set

$$\Phi_n(w) := w - A \log w + \sum_{j+1}^{n} b_j w^{-j}, \tag{2.9}$$

then

$$\Phi_n(F(w)) - \Phi_n(w) - 1 = O(w^{-(n+2)}). \tag{2.10}$$

The logarithm appearing in (2.7), and all other logarithms in this section, are to be understood as the *principal branch*, i.e., the branch with a cut along the negative real axis and real values on the positive axis. Because of the logarithmic term, Φ_{ps} as written is not exactly a formal power series in w^{-1}. To work around this, we rewrite the equation $\Phi_{ps} \circ F = \Phi_{ps} + 1$ formally as

$$F(w) - w - 1 - A \log(F(w)/w) = \sum_{j=1}^{\infty} b_j \left(F(w)^{-j} - w^{-j} \right) \tag{2.11}$$

Since $F(w)/w$ is analytic at ∞ and takes the value 1 there, $\log(F(w)/w)$ is analytic at ∞ and vanishes there. Furthermore, the formal identity

$$\log(F(w)) - \log(w) = \log(F(w)/w)$$

holds literally on $\{-\alpha < \mathrm{Arg}(w) < \alpha, |w| > R\}$ for sufficiently large R, for any $\alpha < \pi$. It is Eq. (2.11) that we really solve.

The left-hand side of (2.11) is analytic at ∞, and vanishes to second order there:

$$F(w) - w - 1 = -Aw^{-1} + O(w^{-2}) \quad \text{and} \quad \log(F(w)/w) = w^{-1} + O(w^{-2})$$

Further, $F(w)^{-j} - w^{-j}$ is analytic at infinity and vanishes to order $j+1$ there, so

$$\sum_{j=1}^{\infty} b_j \left(F(w)^{-j} - w^{-j} \right)$$

is indeed a formal power series in w^{-1} which begins with a term in w^{-2}. Furthermore, the coefficient of w^{-j-1} in the expression on the right in (2.11) can be written as

$$-jb_j + \text{a function of } b_1, \ldots, b_{j-1}.$$

Thus, since the left-hand side of (2.11) is known, the b_j's can be determined successively, and by induction on j, they are uniquely determined. The assertion about the order of the error term $\Phi_n(F(w)) - \Phi_n(w) - 1$ also follows, since

$$\sum_{j=n+1}^{\infty} b_j \left(F(w)^{-j} - w^{-j} \right)$$

begins with a term in $w^{-(n+2)}$.

Theorem 2.2 *Let $b_1, b_2, \ldots$ be as in Proposition 2.13, and let Φ_A be an attracting Fatou coordinate at infinity. Then, for any n,*

$$\Phi_A(w) = w - A \log w + C_A + \sum_{j=1}^{n} b_j w^{-j} + O(|w|^{-(n+1)}) \qquad (2.12)$$

uniformly as $w \to \infty$ in any sector $-\alpha < Arg(w) < \alpha$ with $\alpha < \pi$.

We collect the main estimates needed for the proof of Theorem 2.2 in the following lemma:

Lemma 2.1

$$\sum_{j=0}^{\infty} |F^j(w)|^{-m} = O(|w|^{-(m-1)}) \qquad (I)$$

and

$$\left| (F^j)'(w) \right| \quad \text{is bounded uniformly in } j, \qquad (II)$$

both estimates holding uniformly for $w \to \infty$ in any sector $\{-\alpha < Arg(w) < \alpha\}$ with $\alpha < \pi$.

Proof We fix an $\alpha < \pi$, and we choose an α_1 with $\alpha < \alpha_1 < \pi$; it saves trouble later if we also require that $\pi - \alpha_1 < \pi/6$. Next we fix an R_0 large enough, so that

$$|F(w) - (w + 1)| < \sin(\pi - \alpha_1) \qquad (2.13)$$

holds for $|w| > R_0$; then we fix an R_1 so that the translated sector

$$\Delta(\alpha_1, R_1) := \{w : -\alpha_1 < Arg(w - R_1) < \alpha_1\}$$

does not intersect the disk of radius R_0 about 0. Then (2.13) holds on $\Delta(\alpha_1, R_1)$, so F maps $\Delta(\alpha_1, R_1)$ to itself; also, since $\pi - \alpha_1 < \pi/6$, we also have, again from (2.13),

$$|F(w) - w - 1| < \frac{1}{2} \quad \text{on } \Delta(\alpha_1, R_1), \tag{2.14}$$

from which it follows that

$$\text{Re}(F(w) \geq \text{Re}(w) + 1/2 \quad \text{for } w \in \Delta(\alpha_1, R_1), \tag{2.15}$$

and also

$$- (\pi - \alpha_1) < \text{Arg}(F(w) - w) < (\pi - \alpha_1). \tag{2.16}$$

We will prove estimates (I) and (II) for $w \to \infty$ in $\Delta(\alpha_1, R_1) \cap \{-\alpha < \text{Arg}(w) < \alpha\}$; this does what we want, since every w with $-\alpha < \text{Arg}(w) < \alpha$ and sufficiently large modulus is in $\Delta(\alpha_1, R_1)$.

Changing notation slightly, we want then to estimate

$$\sum_{j=0}^{\infty} |F^j(w_0)|^{-m}$$

for large w_0, given that

$$-\alpha_1 < \text{Arg}(w_0 - R) < \alpha_1 \quad \text{and} \quad -\alpha < \text{Arg}(w_0) < \alpha$$

where, crucially, $\alpha_1 > \alpha$. We write

$$F^j(w_0) =: w_j =: u_j + iv_j.$$

In the calculation which follows, we adopt the convention that K denotes some constant depending only on α, α_1 and m. Different instances of K are not necessarily the same constant. All inequalities involving w_0 are only asserted to hold for $|w_0|$ large enough.

We first treat the case $-\pi/4 \leq \text{Arg}(w_0) \leq \pi/4$, i.e., $|v_0| \leq u_0$. Then, by (2.15),

$$|w_j| \geq u_j \geq u_0 + j/2,$$

so

$$\sum_{j=0}^{\infty} |w_j|^{-m} \leq \sum_{j=0}^{\infty} (u_0 + j/2)^{-m} \leq K u_0^{-(m-1)} \leq K |w_0|^{-(m-1)};$$

in the last step, we used $|v_0| \leq u_0$ to estimate $u_0 \geq |w_0|/\sqrt{2}$. This proves the desired estimate in this case.

There remain the possibilities $\pi/4 < \text{Arg}(w_0) < \alpha$ and $-\alpha < \text{Arg}(w_0) < -\pi/4$. The estimates in the two cases are essentially the same; for definiteness we assume that the first holds, i.e., that w_0 is in the upper half-plane. By (2.16) the w_j are all contained in the translated sector $\{-(\pi - \alpha_1) < w - w_0 < +(\pi - \alpha_1)\}$. This sector intersects the diagonal line $\{\text{Re}(w) = \text{Im}(w)\}$ in a segment $(\hat{w}_-, \hat{w}_+)$ (labeled so that $|\hat{w}_-| < |\hat{w}_+|$). It is easy to show, either by elementary geometry or by writing explicit formulas, that

$$|\hat{w}_-| \geq K^{-1}|w_0| \quad \text{and} \quad |\hat{w}_+| \leq K|w_0|.$$

Since $u_{j+1} \geq u_j + 1/2$, the w_j, which start above the diagonal line, will get to and stay below it after finitely many steps. Let j_0 be the first index for which w_j is below the diagonal. Using the above upper bound on $|\hat{w}_+|$, we get a bound

$$j_0 \leq K|w_0|$$

The line segment from w_{j_0-1} to w_{j_0} intersects the diagonal line between $\hat{w}_-$ and $\hat{w}_+$. Since

$$|w_{j_0} - w_{j_0-1}| = |F(w_{j_0-1}) - w_{j_0-1}| < 3/2,$$

we have

$$|w_{j_0}| \geq |\hat{w}_-| - 3/2 \geq K^{-1}|w_0| - 3/2 \geq K^{-1}|w_0|.$$

By the first case

$$\sum_{j=j_0}^{\infty} |w_j|^{-m} \leq K|w_{j_0}|^{-(m-1)}| \leq K|w_0|^{-(m-1)}. \tag{2.17}$$

All the w_j lie above (or on) the line through w_0 with direction $\alpha_1 - \pi$ (and the origin lies below this line.) Using $\text{Arg}(w_0) < \alpha < \alpha_1$, the distance from this line to the origin satisfies a lower bound $K^{-1}|w_0|$. Hence each $|w_j| \geq K^{-1}|w_0|$. Since $j_0 \leq K|w_0|$,

$$\sum_{j=0}^{j_0-1} |w_j|^{-m} \leq j_0 \cdot \left(K^{-1}|w_0|\right)^{-m} \leq (K|w_0|) \cdot (K^m|w_0|^{-m}) \leq K|w_0|^{-(m-1)}.$$

Combining this estimate on the sum of the first j_0 terms with the estimate (2.17) on the sum of the rest gives

$$\sum_{j=0}^{\infty} |F^j(w_0)|^{-m} \le K |w_0|^{-(m-1)}$$

(for sufficiently large $|w_0|$), so (I) is established. (II) follows easily from (I) together with the estimate

$$F'(w) = 1 + O(|w|^{-2}),$$

the chain rule, and standard manipulations for reducing estimates on products to estimates on sums.

Proof (Theorem 2.2) Let $\pi/2 < \alpha < \pi$. By an argument already used several times, we can choose R sufficiently large so that

$$|F(w) - w - 1| < \sin(\pi - \alpha) \text{ and } \mathrm{Re}(F(w) > \mathrm{Re}\, w + 1/2$$

for all

$$w \in \Delta(\alpha, R) := \{-\alpha < \mathrm{Arg}(w - R) < \alpha\}.$$

As usual, it follows from the first of these inequalities that $F(\overline{\Delta(\alpha, R)}) \subset \Delta(\alpha, R)$. We are going to prove

$$\Phi_A(w) = \Phi_n(w) + O(|w|^{-(n+1)}) \quad \text{as } w \to \infty \text{ in } \Delta(\alpha, R)$$

(where Φ_n is defined by (2.8)); this assertion for all α implies the assertion of the theorem for all α.

We set

$$c_n(w) := \Phi_A(w) - \Phi_n(w) \quad \text{and} \quad u_n(w) := \Phi_n \circ F(w) - \Phi_n(w) - 1.$$

By Proposition 2.13, u_n is analytic at infinity with

$$u_n(w) = d_{n+2} w^{-(n+2)} + \cdots . \tag{2.18}$$

A simple calculation gives

$$c_n \circ F(w) = c_n(w) - u_n(w);$$

iterating gives

$$c_n \circ F^k(w) = c_n(w) - \sum_{j=0}^{k-1} u_n(F^j(w));$$

reorganizing and differentiating gives

$$c_n'(w) = \sum_{j=0}^{k-1} u_n'(F^j(w)) \cdot (F^j)'(w) + c_n'(F^k(w)) \cdot (F^k)'(w). \tag{2.19}$$

By differentiating (2.18)

$$u_n'(w) = O(|w|^{-(n+3)}) \text{ as } w \to \infty,$$

so, by Lemma 2.1,

$$\sum_{j=0}^{\infty} \left| u_n'(F^j(w)) \cdot (F^j)'(w) \right| < \infty.$$

By Proposition 2.5

$$\Phi_A'(w) \to 1 \quad \text{as Re}(w) \to +\infty,$$

and the same is true for Φ_n by an elementary calculation; hence

$$c_n'(w) \to 0 \quad \text{as Re}(w) \to +\infty.$$

Thus, we can let $k \to \infty$ in (2.19) to get

$$c_n'(w) = \sum_{j=0}^{\infty} u_n'(F^j(w)) \cdot (F^j)'(w).$$

Applying both parts of Lemma 2.1 to this representation,

$$|c_n'(w)| \leq \text{const}|w|^{-(n+2)} \tag{2.20}$$

for all $w \in \Delta(\alpha, R)$. It follows from this estimate that the limit

$$\lim_{u \to \infty} c_n(u + iv)$$

exists and is independent of v. We denote this limit by C_A. Then

$$c_n(u + iv) = C_A - \int_u^{\infty} c_n'(\sigma + iv)d\sigma,$$

so by integrating (2.20) we get

$$|c_n(w) - C_A| \le \text{const}|w|^{-(n+1)}$$

for all $w \in \Delta(\alpha, R)$, which is what we set out to prove.   □

We insert here a simple remark which we will want to refer to repeatedly. We say that an analytic function $f : U \to \mathbb{C}$ is *real-symmetric* if the Taylor coefficients of f are real at some point $x \in U \cap \mathbb{R}$. Note, that we do not require that U itself be a real-symmetric domain. Similarly, if x is a point in $\mathbb{R}$, we say that an analytic germ $f(z)$ at x is real-symmetric if its coefficients are real.

Proposition 2.14 *Let f be a real-symmetric analytic germ of the form (2.6), and let ϕ_A be an attracting Fatou coordinate for f. Then there is a pure imaginary constant c so that*

$$\overline{\phi_A(z)} = \phi_A(\overline{z}) + c \quad on \; B_0^f.$$

Furthermore, the coefficients $A, b_1, b_2, \ldots$ of Theorem 2.2 are real.
  Similar assertions hold for a repelling Fatou coordinate.

Proof Let P be a small attracting petal invariant under complex conjugation (e.g., a small disk tangent to the imaginary axis at the origin.) Since f commutes with complex conjugation,

$$z \mapsto \overline{\phi_A(\overline{z})}$$

is another univalent analytic function defined on P and satisfying the usual functional equation $\phi(f(z)) = \phi(z) + 1$. By uniqueness up to an additive constant of the Fatou coordinate (Proposition 2.11), there is a constant c so that

$$\overline{\phi_A(\overline{z})} = \phi_A(z) + c \quad on \; P.$$

In the usual way, this identity extends to all of B_0^f by repeated application of the functional equation. Applying the identity at any real point x shows that c is pure imaginary. We omit the proofs of the other assertions, which are even simpler.   □

2.2.1 A Note on Resurgent Properties of the Asymptotic Expansion of the Fatou Coordinates

Let us briefly mention a very different approach to the construction of the asymptotic series (2.12) originating in the works of Écalle [Ec].

Recall (see e.g., [Ram]) that a formal power series $\sum_{m=1}^{\infty} a_m x^{-m}$ is of *Gevrey order* k if

$$|a_m| < C A^n (n!)^{\frac{1}{k}} \text{ for some choice of positive constants } C, A.$$

Consider the asymptotic expansion (2.7) for the Fatou coordinates, and denote

$$v_*(w) \equiv \sum_{j=1}^{\infty} b_j w^{-j}, \text{ so that } \Phi_{\mathrm{ps}}(w) = w - A \log(w) + v_*(w).$$

As was shown by Écalle for the case $A = 0$ [Ec] and Dudko and Sauzin in the general case [DS1], we have the following.

Theorem 2.3 *The asymptotic series* $v_* = \sum_{j=1}^{\infty} b_j w^{-j}$ *is of Gevrey order* 1.

Recall that the *Borel transform* of a formal power series

$$h_* = \sum_{m=1}^{\infty} a_m x^{-m}$$

consists in applying the termwise inverse Laplace transform:

$$a_m x^{-m} \mapsto \frac{a_m \zeta^{m-1}}{(m-1)!}.$$

In the case when the formal power series is of Gevrey order 1, this yields a series

$$\sum_{m=1}^{\infty} \frac{a_m \zeta^{m-1}}{(m-1)!},$$

which converges to an analytic function $\hat{h}(\zeta)$ in a neighborhood of the origin.

The following theorem describes the phenomenon of *resurgence* associated with the asymptotic series v_*, discovered by Écalle [Ec]. Écalle presented the proof for the case when $A = 0$, and so the logarithmic term is absent in (2.7), and he outlined an approach to it in the general case. An independent proof in the general case was recently given by Dudko and Sauzin [DS1].

Theorem 2.4 *The Borel transform* $\hat{v}$ *of the formal power series* v_* *analytically extends from the neighborhood of the origin along every path which avoids the points* $2\pi i \mathbb{Z}^*$. *Furthermore, let* $S_{\theta, \varepsilon}$ *be any sector*

$$S_{\theta, \varepsilon} = \{|Arg(\zeta) - \theta| < \varepsilon\} \text{ such that } \overline{S_{\theta, \varepsilon}} \cap \{2\pi i \mathbb{Z}^*\} = \emptyset,$$

and let γ *be any path as above which eventually lies in* $S_{\theta, eps}$ *(see Fig. 2.1). Denote* $\hat{v}^{\gamma}$ *the analytic continuation along* γ. *Then* $\hat{v}^{\gamma}$ *is a function of exponential type:*

Fig. 2.1 An analytic
continuation along a path γ
has an exponential type in a
sector $S_{\theta,\varepsilon}$ whose closure
does not contain any of the
points $2\pi i \mathbb{Z}^*$

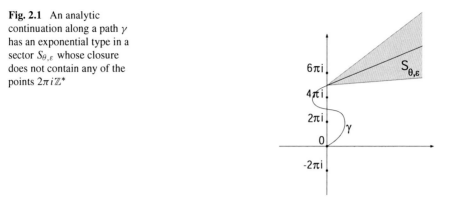

$$|\hat{h}^{\gamma}(\zeta)| < C \exp(D|\zeta|),$$

*where the constant C and D depend only on θ and ε. In particular, $\hat{h}$ has an analytic
continuation $\hat{h}^+$ to the right half-plane {Re ζ > 0} and an analytic continuation $\hat{h}^-$
to the left half-plane {Re ζ < 0}.*

Consider the standard Laplace transforms

$$\mathcal{L}^+(w) = \int_0^\infty e^{-w\zeta} \hat{h}(\zeta)d\zeta \text{ and } \mathcal{L}^-(w) = \int_{-\infty}^0 e^{-w\zeta} \hat{h}(\zeta)d\zeta.$$

Note that

$$v^+ \equiv \mathcal{L}^+\hat{v}^+ \text{ and } v^- \equiv \mathcal{L}^-\hat{v}^-$$

are defined for Re w sufficiently large. The resurgent properties of v_* are completed
by the following refinement of Theorem 2.2:

Theorem 2.5 [DS1] *The analytic functions*

$$\Phi_A(w) = w - A\log w + v^+(w) \text{ and } \Phi_R(w) = w - A\log w + v^-(w).$$

*Furthermore, let $\alpha < \pi/2$. Then there exist positive constants $B = B(\alpha)$, $C = C(\alpha)$
such that*

$$\left| \Phi_A(w) - \left(w - A\log w + \sum_{j=1}^n b_j w^{-j} \right) \right| < C^{n+1}(n+1)!|w|^{-(n+1)}$$

uniformly in a sector {$|Arg(w)| < \pi/2 + \alpha$ and $|w| > B$}, and similarly for Φ_R.

It follows from Theorem 2.5 and a Stirling formula estimate that the first n terms of the asymptotic series v_* are useful for numerically estimating the Fatou coordinates for $|w| > \text{const} \cdot n$.

The values of the constants in Theorem 2.5 can be estimated explicitly. Dudko [Du] shows the following. Let us write

$$F(w) = w + 1 + a(w), \quad \text{where } a(w) = Aw^{-1} + O(w^{-2}),$$

and introduce functions

$$b(w) = a(w - 1) = \sum_{k=1}^{\infty} c_k w^{-k}, \quad \text{and}$$

$$m(w) = -A \log \left(\frac{1 + w^{-1} b(w)}{1 - w^{-1}} \right) + b(w) = \sum_{k=1}^{\infty} d_k w^{-k}.$$

Let $C_0, \beta > 0$ be such that for all $k \in \mathbb{N}$,

$$|c_k| \le C_0 \beta^{k-1} \quad \text{and} \quad |d_k| \le C_0 \beta^{k-1}.$$

Finally, let S be such that $|\hat{v}(\zeta)| \le S$ for $|\zeta| \le 2$. Set

$$B = \frac{(\beta_0 + \frac{C_0}{\cos \alpha})(\frac{\alpha}{8} + 1) + 1}{\sin \frac{\alpha}{8}}.$$

Then for all $n \in \mathbb{N}$

$$\left| \Phi_A(w) - \left(w - A \log w + \sum_{j=1}^{n} b_j w^{-j} \right) \right| < \left(S + \frac{C_0}{\cos \alpha} \left(\frac{8}{\alpha} \right)^{n+1} \right) (n + 1)! |w|^{-(n+1)}$$

uniformly in a sector $\{|\text{Arg}(w)| < \pi/2 + \alpha$ and $|w| > B\}$, and similarly for Φ_R.

2.3 Écalle-Voronin Invariants and Definition of Parabolic Renormalization

The Riemann surface $\mathbb{C}/\mathbb{Z}$ has two punctures at the upper end ($\text{Im } z \to +\infty$), and at the lower end ($\text{Im } z \to -\infty$). Filling them with points $\oplus$ and $\ominus$ respectively, we obtain the Riemann sphere. The mapping

$$\text{ixp}(z) \equiv \exp(2\pi i z),$$

conformally transforms $\mathbb{C}/\mathbb{Z} \mapsto \mathbb{C}^*$, sending $\oplus \to 0$ and $\ominus \to \infty$.

Consider a germ f with a simple parabolic fixed point at 0, normalized as in (2.6). Let P_A and P_R be a pair of ample petals for f, and denote f^{-1} the local branch of the inverse which fixes the origin. Note that f^{-1} extends univalently to $P_R \cup f(P_A)$. Fix a choice of the Fatou coordinates ϕ_A and ϕ_R.

The forward orbits originating in P_A are parametrized by points in the attracting cylinder $\mathscr{C}_A$. Similarly, f^{-1}-orbits in P_R are parametrized by points in $\mathscr{C}_R$. By the definition of an ample petal, $P_A \cap P_R \neq \emptyset$. Let z be any point in the intersection of the petals. It is trivial to see that the correspondence

$$\tilde{\phi}_R(z) \mapsto \tilde{\phi}_A(z)$$

defines a mapping from a subset of $\mathscr{C}_R$ to $\mathscr{C}_A$. We denote this mapping by h. It is more convenient for us to pass to $\mathbb{C}^*$ from $\mathbb{C}/\mathbb{Z}$ via the exponential, and consider the mapping $\mathbf{h}$ formally defined as

$$\mathbf{h} = (\mathrm{ixp} \circ \phi_A) \circ (\mathrm{ixp} \circ \phi_R)^{-1}. \tag{2.21}$$

We note:

Lemma 2.2 *The mapping $\mathbf{h}$ is analytic.*

Furthermore,

Lemma 2.3 *The domain of definition of $\mathbf{h}$ contains a punctured neighborhood of 0, and a punctured neighborhood at infinity. The singularities of $\mathbf{h}$ at 0 are removable, the analytic extension taking the value 0 at 0. Similarly, $\mathbf{h}$ has a pole at ∞.*

Proof By local theory, $P_A \cap P_R$ contains the (upward-facing) circular sector

$$\{w : 0 < |w| < r, \pi/2 - \delta < \mathrm{Arg}(w) < \pi/2 + \delta\}$$

for sufficiently small r, for any $\delta < \pi/2$. By Proposition 2.6 and the crude asymptotic estimate $\phi_R(z) \approx -1/z$ as $z \to 0$ staying away from the positive real axis (Proposition 2.5), it follows that the image under ϕ_R of this sector contains the upper half-strip

$$\phi_R(P_A \cap P_R) \supset \{x + iy : 0 \leq x \leq 1, y > R\} \tag{2.22}$$

for sufficiently large R, so the domain of definition of $\mathbf{h}$ contains a punctured neighborhood

$$\{z : 0 < |z| < \exp(-2\pi R)\}.$$

By Proposition 2.5 the infimum of $\mathrm{Im}(\phi_A(x + iy))$ over the strip on the right of (2.22) goes to $+\infty$ as $R \to \infty$, so $\mathbf{h}(z) \to 0$ as $z \to 0$, as claimed. The proofs for the assertions about ∞ are similar. ☐

Let us analytically continue $\mathbf{h}$ to the origin and to infinity. We denote $h^+(z)$ the analytic germ of $\mathbf{h}$ at 0, and $h^-(z)$ the analytic germ at ∞. When necessary to emphasize the dependence on the germ f we will write $\mathbf{h}_f$ and $h_f^{\pm}$.

It is easy to see the following.

Proposition 2.15 *The germs h^+, h^- do not depend on the choice of petals P_A and P_R.*

By Proposition 2.11, an attracting (repelling) Fatou coordinate differs from our choice of ϕ_A (ϕ_R) by an additive constant. A trivial verification shows that replacing ϕ_A by $\phi_A + c_A$ and ϕ_R by $\phi_R + c_R$ changes $h^{\pm}$ to

$$w \mapsto \lambda_A h^{\pm}(\lambda_R^{-1} w) \tag{2.23}$$

with

$$\lambda_A = \exp(2\pi i c_A), \quad \lambda_R = \exp(2\pi i c_R). \tag{2.24}$$

The scale change factors λ_A and λ_R can be given arbitrary nonzero values by the appropriate choices of c_R and c_A.

We are going to show that the zero of h^+ at 0 and the pole of h^- at ∞ are simple. We will see this by deriving a useful explicit formula for the respective leading coefficients. To write this formula, we need to introduce some notation. Writing just the first few terms in the asymptotic approximations to the Fatou coordinates (Proposition 2.2):

$$\begin{aligned}
\phi_A(z) &= -\frac{1}{z} + A\log(-z) + C_A + O(z) \\
\phi_R(z) &= -\frac{1}{z} + A\log(z) + C_R + O(z)
\end{aligned} \tag{2.25}$$

where

- $A = 1 - \alpha$, with α the coefficient of z^3 in the Taylor expansion for f at 0;
- the logarithms mean the *standard branch*, i.e., the branch with a cut along the negative real axis and real values on the positive real axis;
- C_R and C_A are complex constants (specifying the normalization of the Fatou coordinates);
- the term $O(z)$ in the first equation means a quantity that goes to zero at least as fast as $|z|$ as $z \to 0$ inside any sector of the form $\{z \neq 0 : -\alpha < \text{Arg}(z) < \alpha\}$ with $\alpha < \pi$ (a *left-facing* sector), and $O(z)$ in the second means as $z \to 0$ in any similarly defined right-facing sector.

Proposition 2.16

$$(h^+)'(0) = \exp\left(-2A\pi^2 + 2\pi i(C_A - C_R)\right)$$

and

$$\lim_{z \to \infty} \frac{h^-(z)}{z} = \exp\big(+2A\pi^2 + 2\pi i(C_A - C_R)\big).$$

In particular, h^+ has a simple zero at 0, h^- a simple pole at ∞, and

$$(h^+)'(0) \cdot (h^-)'(\infty) = \exp(-4\pi^2 A). \tag{2.26}$$

Proof To prove the formula for $(h^+)'(0)$, we look at points $z(t)$ of the form it, with t small and positive; such w's are in $B_0^f \cap P_R$ for any ample petal P_R. Inserting into (2.25) and using $\log(\pm it) = \log(t) \pm i\pi/2$,

$$\phi_R(z(t)) = it^{-1} + A \log t + iA\pi/2 + C_R + O(t)$$

$$\phi_A(z(t)) = it^{-1} + A \log t - iA\pi/2 + C_A + O(t).$$

We put $w(t) := \exp(2\pi i \phi_R(z(t)))$; then $h^+(w(t)) = \exp(2\pi i \phi_A(z(t)))$, so

$$\frac{h^+(w(t))}{w(t)} = \exp(2\pi i(-iA\pi + C_A - C_R + O(t))),$$

and so

$$(h^+)'(0) = \lim_{t \to 0^+} \frac{h^+(w(t))}{w(t)} = \exp\big(-2A\pi^2 + 2\pi i(C_A - C_R)\big),$$

as asserted. The assertion about h^- is proved by a similar calculation. □

Let us say that two pairs of germs at zero and infinity (h_1^+, h_1^-) and (h_2^+, h_2^-) are *equivalent* if there exist nonzero constants λ_A and λ_R so that

$$h_2^\pm(z) = \lambda_A h_1^\pm(\lambda_R^{-1} z).$$

In view of the following, let us call the equivalence class of the germs $\mathbf{h}_f = h_f^\pm$ the *Écalle-Voronin invariant* of f. By (2.26) the Écalle-Voronin invariant determines the *formal* conjugacy class of the simple parabolic germ f. A much stronger result is due to Voronin [Vor] (an equivalent version was formulated by Écalle [Ec]).

Theorem 2.6 *Two analytic germs f_1 and f_2 of the form (2.6) are conjugate by a local conformal change of coordinates $\varphi(z)$ with $\varphi(0) = 0$ if and only if their Écalle-Voronin invariants are equal.*

Sketch of proof The proof of the "only if" direction is very easy—essentially a diagram-chase. We assume

$$f_1 = \varphi^{-1} \circ f_2 \circ \varphi \quad \text{on a neighborhood of } 0,$$

where φ is analytic at 0 with $\varphi(0) = 0$, $\varphi'(0) > 0$. We label various objects attached to f_1 and f_2 with indices 1 and 2 respectively. Because the assertion concerns germs, we can cut down the domains of the various functions appearing as convenient. We set things up as follows:

- We fix a domain for φ so that it is univalent.
- We fix a domain for f_2 contained in the image of φ, on which f_2 is univalent, and so that the image of f_2 is contained in the image of φ.
- We take for the domain of f_1 the preimage under φ of the domain of f_2. Then the equation

$$f_1 = \varphi^{-1} \circ f_2 \circ \varphi$$

is exact, including domains.

It is then obvious that:

- φ maps any ample petal for f_1 to an ample petal for f_2. We fix any ample repelling petal $P_R^{(1)}$ to use in the construction of $\mathbf{h}_1$, and we use $P_R^{(2)} := \varphi P_R^{(1)}$ as a repelling petal to construct $\mathbf{h}_2$.
- If $\phi_A^{(2)}$ is an attracting Fatou coordinate for f_2 defined on B^{f_2}, then $\phi_A^{(2)} \circ \varphi$ is an attracting Fatou coordinate for f_1, and similarly for repelling Fatou coordinates. To construct Écalle-Voronin pairs for the two mappings, we choose any attracting Fatou coordinate $\phi_A^{(2)}$ for f_2 and use $\phi_A^{(1)} := \phi_A^{(2)} \circ \varphi$ for f_1 (and similarly for repelling Fatou coordinates).

With things organized this way, an entirely mechanical verification shows that $\mathbf{h}_1$ and $\mathbf{h}_2$ are identical pairs of germs.

Conversely, assume that the Écalle-Voronin invariants of f_1 and f_2 are equal. By fixing the constants in Fatou coordinates for the two mapping appropriately, we can arrange that

$$h_1^+ = h_2^+ \quad \text{and} \quad h_1^- = h_2^- \tag{2.27}$$

on neighborhoods of 0 and ∞ respectively. Fix also small attracting and repelling petals, for instance,

$$P_R = -P_A = \{z : -1/z \in \Delta(\alpha, R)\},$$

with $\Delta(\alpha, R)$ as defined in (2.3), α some number in $(\pi/2, \pi)$, and R large enough. With these choices, $P_R \cap P_A$ has only two components, an upper one contained in U^+ and a lower one contained in U^-. Then $\left(\phi_A^{(2)}\right)^{-1} \circ \left(\phi_A^{(1)}\right)$ conjugates f_1 to f_2 on P_A, and correspondingly, $\left(\phi_R^{(2)}\right)^{-1} \circ \left(\phi_R^{(1)}\right)$ conjugates f_1 to f_2 on P_R. From (2.27) it follows, possibly after adding appropriate integers to the various Fatou coordinates, that the two conjugators agree on $P_A \cap P_R$. Putting them together, we obtain a conjugator on $P_A \cup P_R$, a punctured neighborhood of 0. It is easy to see, using the asymptotic estimates for Fatou coordinates, that this conjugator extends analytically through 0 with the value 0 there.   □

In fact, all equivalence classes of pairs of germs actually occur as Écalle-Voronin invariants.

Theorem 2.7 *Let h^+ be an analytic germ at 0, with a simple zero there, and let h^- be a meromorphic germ at ∞ with a simple pole there. Denote $\lambda^+ = (h^+)'(0)$, $\lambda^- = (h^-)'(\infty)$, and let*

$$\lambda^+ \cdot \lambda^- = e^{-4\pi^2 A}.$$

Then there exists a simple parabolic germ $f(z)$ at the origin of the form

$$f(z) = z + z^2 + (1 - A)z^3 + \cdots$$

whose Écalle-Voronin invariant is the equivalence class of (h^+, h^-).

Sketch of proof We follow the argument given in [BH]. Let us choose the lifts $H^\pm(w)$ of the germs $h^\pm$ via the exponential map

$$e^{2\pi i H^\pm(w)} = h^\pm(e^{2\pi i w}).$$

For a sufficiently large value of $R > 0$, the maps $H^\pm$ are defined in the half-planes $V^\pm \equiv \{\pm \operatorname{Im} w > R\}$. Denote $U^\pm \equiv H^\pm(V^\pm)$—these domains are invariant under the translation $w \mapsto w + 1$ and contain half-planes $\{\pm \operatorname{Im} w > R'\}$. We have

$$H^\pm(w) = w + \frac{1}{2\pi i} \log(\lambda^\pm) + o(1);$$

let us select the branches of the logarithm so that $\log(\lambda^+ \lambda^-) = 4\pi^2 A$. Fix $w^\pm \equiv H^\pm(R \pm i R)$. We define V as the union of $V^\pm$ and the left half-plane $\{\operatorname{Re}(w) < -R\}$ and U as the union of $U^\pm$ and the half-plane to the right of the line passing through w^+ and w^-. Let $\mathcal{V}$ denote the Riemann surface obtained by gluing V and U via H^+, H^- (see Fig. 2.2). The result follows from the following claim:

Claim *For sufficiently large values of R, the Riemann surface $\mathcal{V}$ is conformally isomorphic to $\mathbb{C} \setminus \mathbb{D}$.*

The proof of the claim is fairly straightforward, so we leave it to the reader. Let us show how to complete the argument assuming the claim. Let us select a large enough value of R, and let $\chi : \mathcal{V} \to \mathbb{C} \setminus \mathbb{D}$ be the conformal isomorphism with $\chi'(\infty) > 0$. Denote

$$V' = V \setminus \{-R - 1 \leq \operatorname{Re}(w) < -R\},$$

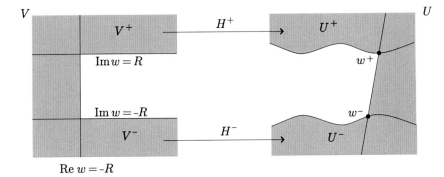

Fig. 2.2 An illustration of the proof of Theorem 2.7

and let $\mathscr{V}'$ be a subsurface of $\mathscr{V}$ obtained by replacing V with V'. The unit translation $w \mapsto w + 1$ maps $V' \to V$ and $U \to U$ and commutes with $H^{\pm}$. Hence it induces an analytic map $F : \mathscr{V}' \to \mathscr{V}$. It is easy to see that

$$\hat{F}(w) \equiv \chi \circ F \circ \chi^{-1}(w) = w + 1 + \frac{A}{w} + o(\frac{1}{w}).$$

Indeed, the projections to V and U are the Fatou coordinates at infinity for $\hat{F}$, and $H^{\pm}$ are the changes of coordinates between V and U. By construction,

$$f(z) \equiv -\frac{1}{\hat{F}(-\frac{1}{z})}$$

is an analytic germ at the origin with the desired properties.   □

We are finally ready to define parabolic renormalization:

Definition 2.2 We say that a parabolic germ $f(z)$ of the form (2.2) is *renormaliz-able* if for some—and hence for all—choices of normalization of ϕ_A and ϕ_R, the coefficient of z^2 in the Taylor series of h^+ at 0 does not vanish.

Then, by the rescaling formulas (2.23) and (2.24), there exist choices of the normalizations of the attracting and repelling Fatou coordinates so that the corresponding h^+ has the form

$$h^+(w) = w + w^2 + \cdots \tag{2.28}$$

i.e., so that h^+ has a normalized simple parabolic fixed point at 0. The rescaling factors λ_A and λ_R that accomplish this are uniquely determined; the corresponding additive constants c_A and c_R are uniquely determined modulo $\mathbb{Z}$.

Definition 2.3 For a renormalizable germ $f(z)$ of the form (2.6), we will call the analytic germ of the unique rescaling of h^+ with the form (2.28) the *parabolic renormalization* of f. We will use the notation $\mathscr{P}f$ for the parabolic renormalization.

We thus have

$$\mathscr{P}f = \mathrm{ixp} \circ \phi_A \circ (\phi_R)^{-1} \circ \mathrm{ixp}^{-1}, \qquad (2.29)$$

with ϕ_A, ϕ_R the appropriately normalized Fatou coordinates and with suitably selected branches of the inverses.

2.4 Analytic Continuation of Parabolic Renormalization

Parabolic renormalization, as defined above, maps a renormalizable analytic germ f at the origin of the form (2.6) to a germ $\mathscr{P}(f)$ of the same form. We will change the point of view now, and talk about an analytic map f, defined in a domain $\mathrm{Dom}(f) \ni 0$, whose germ at 0 is of the form (2.6). Note that we do not impose any conditions on the naturality of the domain $\mathrm{Dom}(f)$ at this point. The map f may analytically extend beyond $\mathrm{Dom}(f)$, however, when considering orbits of points under f, we restrict ourselves only to the orbits which do not leave $\mathrm{Dom}(f)$.

Here, and elsewhere in the book, for a map f with a *natural* domain of definition we use the notation $\mathscr{D}(f)$ instead of $\mathrm{Dom}(f)$.

As before, we denote $B^f \subset \mathrm{Dom}(f)$ the basin of 0, and $B_0^f \subset B^f$ the immediate basin of 0, that is, the connected component of B^f which contains an attracting petal. Let us fix an attracting Fatou coordinate ϕ_A and extend it to all of B^f via the functional equation. We also choose a repelling petal P_R and fix a repelling Fatou coordinate ϕ_R on it.

We make the following simple observation:

Lemma 2.4 *Let $z_1 \in B^f \cap P_R$ and let $z_2 \in P_R$, and assume*

$$\mathrm{ixp}(\phi_R(z_1)) = \mathrm{ixp}(\phi_R(z_2)).$$

Then $z_2 \in B^f$ and

$$\mathrm{ixp}(\phi_A(z_1)) = \mathrm{ixp}(\phi_A(z_2)).$$

Proof From the assumption,

$$\phi_R(z_2) - \phi_R(z_1) =: m \in \mathbb{Z}.$$

We consider separately $m \geq 0$ and $m < 0$. In the first case, $\phi_R(f^{-m}(z_2)) = \phi_R(z_1)$. Since $f^{-m}(z_2)$ and z_1 are both in P_R, and since ϕ_R is univalent on P_R, $f^{-m}(z_2) = z_1$, hence $z_2 = f^m(z_1)$, so $z_2 \in B^f$ and $\phi_A(z_2) = \phi_A(z_1) + m$ and the asserted equality holds.

Now suppose $m < 0$, and let $p := -m > 0$. Then $\phi_R((f^{-p}(z_1) = \phi_R(z_2)$, and, arguing as above, $z_1 = f^p(z_2)$, so $z_2 \in B^f$ and, again, the asserted equality holds. □

The above lemma implies the following.

Corollary 2.1 *The pair of analytic germs* **h** *extends to*

$$\mathscr{D}(\mathbf{h}) \equiv \{w = \mathrm{ixp} \circ \phi_R(z) \mid \text{where } z \in B^f \cap P_R\} \cup \{0, \infty\}.$$

Furthermore, we have the following.

Proposition 2.17 *The domain* $\mathscr{D}(h)$ *does not depend on the choice of the repelling petal* P_R.

Proof Suppose $P_R^{(1)}$ is another repelling petal, and write $\mathbf{h}^{(1)}$ for the corresponding function. If $w \in \mathscr{D}(\mathbf{h})$, then w can be written as $\mathrm{ixp} \circ \phi_R(z)$, with $z \in B^f \cap P_R$. For large enough n, $f^{-n}(z) \in P_R^{(1)}$, and, since $f^n(f^{-n}(z)) = z \in B^f$, then $f^{-n}(z) \in B^f$. Hence $w = \mathrm{ixp} \circ \phi_R(f^{-n}(z)) \in \mathscr{D}(\mathbf{h}^{(1)})$ and

$$\mathbf{h}^{(1)}(w) = \mathrm{ixp} \circ \phi_A(f^{-n}(z)) = \mathrm{ixp} \circ \phi_A(z) = \mathbf{h}(w).$$

This shows that $\mathscr{D}(\mathbf{h}) = \mathscr{D}(\mathbf{h}^1)$. □

Since the derivative of the local inverse never vanishes, we have:

Lemma 2.5 *The only possibilities for critical values of* **h** *in* $\mathscr{D}(\mathbf{h})$ *are the images under* ixp *of the critical values of* ϕ_A. *More explicitly:* v *is a critical value of* **h** *if and only if it can be written*

$$v = \mathrm{ixp} \circ \phi_A(z_c)$$

where z_c *is a critical point of* f *belonging to* B^f *and admitting a backward orbit converging to* 0.

We further prove:

Theorem 2.8 *Suppose* B_0^f *is a Jordan domain, such that* ϕ_A *cannot be analytically continued through any point of* ∂B_0^f. *Then the pair of germs* $\mathbf{h} = (h^+, h^-)$ *has a maximal domain of analyticity* $\mathscr{D}(\mathbf{h})$, *which is a union of two Jordan domains* $W^+ \ni 0$ *and* $W^- \ni \infty$. *Furthermore, let* P_R *be a repelling petal for* f. *Then*

$$\partial \mathscr{D}(\mathbf{h}) \subset \mathrm{ixp} \circ \phi_R(\partial B_0^f \cap P_R).$$

Proof Let P_R be a repelling petal which maps under ϕ_R to a left half-plane. After shrinking P_R if necessary, we further assume that $f(P_R)$ is also a petal. For purposes of this proof, f^{-1} and ϕ_R^{-1} will mean the (two-sided) inverses of the respective restrictions to P_R. By local theory, $P_R \cup \{0\}$ contains a neighborhood of 0 in ∂B_0^f. The component of $\partial B_0^f \cap (P_R \cup \{0\})$ containing 0 is therefore a Jordan arc, which we denote by $\tilde{\sigma}$. The intersection $\overline{P_R} \cap (\partial B_0^f \setminus \tilde{\sigma})$ is compact and does not contain 0, so $\mathrm{Re}(\phi_R(\,.\,))$ is bounded below on it. Hence, for β sufficiently negative, the arc

$$\tilde{\gamma} := \phi_R^{-1}\{\mathrm{Re}\, w = \beta\}$$

is disjoint from $\partial B_0^f \setminus \tilde{\sigma}$. We give the arc $\tilde{\gamma}$ the clockwise orientation, which means that $\mathrm{Im}(\phi_R(\,.\,))$ goes to $+\infty$ at the beginning of $\tilde{\gamma}$. The arc $\tilde{\gamma}$ must intersect B_0^f; otherwise, it (with 0 appended) would bound a repelling petal contained in B_0^f, which is impossible by local theory. Thus

$$\emptyset \neq \tilde{\gamma} \cap \partial B_0^f \subset \tilde{\sigma}.$$

Let γ denote the initial segment of $\tilde{\gamma}$ up to its first intersection with B_0^f, and denote the first intersection point by z_1. We have set things up so as to guarantee that $z_1 \in \tilde{\sigma}$. Then $f^{-1}\gamma$ is another arc in $B_0^f \cap P_R$, running from 0 to $f^{-1}(z_1)$, and disjoint from γ. Let σ denote the subarc of ∂B_0^f running from z_1 to $f^{-1}(z_1)$ (with end-points included this time). Then

$$\delta := \gamma + \sigma - f^{-1}\gamma$$

(meaning: first traverse γ, then σ, then $f^{-1}\gamma$ backwards) is a Jordan curve. Since $\delta \setminus \{0\}$ is contained in the Jordan domain P_R, the domain Δ bounded by δ is also contained in P_R. Since $\overline{P_R}$ is contained in the petal $f(P_R)$, the Fatou coordinate ϕ_R is analytic on $\overline{\Delta} \setminus \{0\}$, and

$$W := \mathrm{ixp} \circ \phi_R(\Delta \cup \gamma \cup f^{-1}\gamma) \tag{2.30}$$

is contained in $\mathscr{D}(\mathbf{h})$. We will argue that $W \cup \{0\}$ is a Jordan domain, and that it is equal to W^+. The main steps are to show the following.

1. W is a connected open punctured neighborhood of 0.
2. $\mathrm{ixp} \circ \phi_R$ maps σ to a Jordan curve $\hat{\sigma}$.
3. The boundary of W is $\hat{\sigma} \cup \{0\}$.

Since W is the image under $\mathrm{ixp} \circ \phi_R$ of a subset of $P_R \cap B_0^f$, $W \subset \mathscr{D}(\mathbf{h})$. On the other hand, $\hat{\sigma}$ is contained in the image of ∂B_0^f and hence disjoint from $\mathrm{Dom}(f)$. By the Jordan curve theorem, $\mathbb{C} \setminus \hat{\sigma}$ has exactly two connected components. As $W \cup \{0\}$ is connected, open, and disjoint from $\hat{\sigma}$, it is contained in one of these components; we temporarily denote the containing component U. If W were not all of U, then its relative boundary in U would have to be nonempty, contradicting (3). Thus, $W \cup \{0\}$ is a Jordan domain. Further, $W \subset \mathscr{D}(\mathbf{h})$ and $\hat{\sigma}$ does not intersect $\mathscr{D}(\mathbf{h})$, so W must be a component of $\mathscr{D}(\mathbf{h})$, i.e., must be $W^+ \setminus \{0\}$.

To prove (1): By definition (2.30) is the continuous image of a connected set and so connected. The image of $\Delta \cup \gamma$ under ϕ_R contains an upper half-strip

$$\{u + iv : \beta - 1 < u \le \beta, v > R\}$$

for a sufficiently large R, and this half-strip maps under ixp to a punctured disk about 0, so W is a punctured neighborhood of 0.

To show that W is open, it suffices to show that the image of γ under $\mathrm{ixp} \circ \phi_R$, which is the same as the image of $f^{-1}\gamma$, is in the interior of W. This follows from the way $\mathrm{ixp} \circ \phi_R$ glues together the two edges. Roughly, any sufficiently small disk about a point of the image of γ is the disjoint union of three parts, respectively the images under $\mathrm{ixp} \circ \phi_R$ of

- a differentiably-distorted half-disk in Δ with diameter along γ,
- another differentiably-distorted half-disk in Δ with diameter along $f^{-1}\gamma$,
- a subarc of γ.

Thus, any such disk is contained in the interior of W.

To prove (2), we need

Lemma 2.6 $ixp \circ \phi_R$ *is injective on* $\sigma \setminus \{z_1, f^{-1}(z_1)\}$ *(that is, on* σ *with its end-points deleted).*

We postpone the proof and proceed to deduce (2) from the lemma. Composing $\mathrm{ixp} \circ \phi_R$ with a continuous parametrization of σ gives a continuous mapping from the parameter interval $[0, 1]$ to $\mathbb{C}$ which is injective except for sending 0 and 1 to the same point. In an obvious way, this produces a continuous injective mapping of the circle $\mathbb{T}$ to $\mathbb{C}$, that is, a parametrized Jordan curve in $\mathbb{C}$.

To prove (3): it is nearly obvious that $\hat{\sigma}$ and 0 are contained in the boundary of W. To prove the converse, let w be a boundary point of W; let (w_n) be a sequence in W converging to w; and, for each n, let z_n be a point of $\Delta \cup \gamma$ with

$$w_n = \mathrm{ixp} \circ \phi_R(z_n).$$

By compactness of $\overline{\Delta}$, we can assume, by passing to a subsequence, that

$$z_n \to z \in \overline{\Delta}.$$

If $z = 0$, then $z_n \to 0$ inside $\Delta \cup \gamma$, which implies

$$\mathrm{Im}(\phi_R(z_n)) \to \infty,$$

which implies

$$w_n \to 0, \text{ i.e., } w = 0.$$

Otherwise, $w = \text{ixp} \circ \phi_R(z)$. Since $w \notin W$, z cannot be in Δ, or in γ, or in $f^{-1}\gamma$, and we have already dealt with the possibility $z = 0$, so we are left only with $z \in \sigma$, which implies

$$w = \text{ixp} \circ \phi_R(z) \in \widehat{\sigma}.$$

Modulo the proof of Lemma 2.6, this shows that W^+ is a Jordan domain and also gives a useful representation for ∂W as the image under $\text{ixp} \circ \phi_R$ of a fundamental domain for f in ∂B_0^f. From this latter representation, it is evident that if $\text{ixp} \circ \phi_A$ cannot be analytically continued through ∂B_0^f, then $h^+ = h|_{W^+}$ cannot be analytically continued through ∂W. Thus, all the assertions about h^+ are proved; the proofs of those about h^- are similar.   □

Proof (Lemma 2.6) We use the same notation as in the proof of Proposition 2.8. Recall that $\widetilde{\sigma}$ denotes the component of $B_0^f \cap P_R$ which contains the parabolic point 0. Deleting 0 splits $\widetilde{\sigma}$ into two subarcs, denoted by $\widetilde{\sigma}_\pm$; we will say later which is which. We parametrize $\widetilde{\sigma}$ as $t \mapsto \widetilde{\sigma}(t) : t_- < t < t_+$, with parameter $t = 0$ corresponding to the parabolic point 0, and with $\widetilde{\sigma}_+$ corresponding to $t > 0$. Since f^{-1} maps P_R into itself and $\widetilde{\sigma}$ into ∂B_0^f, it maps $\widetilde{\sigma}$ into itself. Using the parametrization, we conjugate f^{-1} on $\widetilde{\sigma}$ to a one-dimensional mapping

$$f^{-1}(\widetilde{\sigma}(t)) = \widetilde{\sigma}(j(t)),$$

where $j(.)$ is a continuous injective mapping of the parameter interval into itself, with $j(0) = 0$.

Claim $j(.)$ *is increasing, i.e., (loosely) f is orientation-preserving on* B_0^f.

We assume the claim for the moment. Since f^{-n} converges uniformly to 0 on P_R,

$$0 < j(t) < t \quad \text{for } 0 < t < t_+.$$

Recall that $\widetilde{\gamma}$ is the image under ϕ_R^{-1} of an appropriate vertical line and that γ is the initial segment of $\widetilde{\gamma}$, up to z_1, its first point of intersection with B_0^f. We choose the labelling of the components of $\widetilde{\sigma} \setminus \{0\}$ so that $z_1 \in \widehat{\sigma}_+$. Then $f^{-1}(z_1)$ is also in $\widetilde{\sigma}_1$ and, from the conjugacy to $j(.)$, the subarc of $\widetilde{\sigma}$ from z_1 to $f^{-1}(z_1)$ with one endpoint included and the other not is a fundamental domain for the action of f^{-1} on $\widetilde{\sigma}_+$. In particular, two distinct points on this arc have disjoint f^{-1} orbits and hence distinct images under $\text{ixp} \circ \phi_R$.

It remains to prove the claim. We know that f^{-1} maps $\widetilde{\sigma}$ into itself and hence maps $\widetilde{\sigma}_+$ either into itself, or into $\widetilde{\sigma}_-$. From injectivity of f^{-1} on $\widetilde{\sigma}$, we will be done if we show that the first alternative holds:

$$f^{-1}\tilde{\sigma}_+ \subset \tilde{\sigma}_+.$$

By connectivity, this will follow if we show that $f^{-1}\tilde{\sigma}_+ \cap \tilde{\sigma}_+$ is nonempty.

The labelling of the components of $\tilde{\sigma} \setminus \{0\}$ is chosen by requiring that $\tilde{\gamma}$ meet $\tilde{\sigma}_+$ before $\tilde{\sigma}_-$. The closed path made by following $\tilde{\gamma}$ from 0 to its first meeting point z_1 with $\tilde{\sigma}$, then $\tilde{\sigma}_+$ back to 0, is a Jordan curve; denote the domain it bounds by U. The first place $\tilde{\sigma}_-$ meets $\tilde{\gamma}$ is outside $\overline{U}$; since $\tilde{\sigma}_-$ does not intersect the boundary of U, all of $\tilde{\sigma}_-$ is outside of U. Thus, any continuous path in P_R which starts in U and reaches $\tilde{\sigma}_-$ must intersect $\tilde{\sigma}_+$ first. It is easy to see, using local theory, that $f^{-1}\gamma$ starts out in U. The first place where $f^{-1}\gamma$ intersects $\tilde{\sigma}$ is $f^{-1}(z_1)$, so

$$f^{-1}(z_1) \in \tilde{\sigma}_+ \cap f^{-1}\tilde{\sigma}_+,$$

completing the proof.   □

Let us introduce a model for the dynamics of a map on its immediate basin. We use the notation $B : \mathbb{D} \to \mathbb{D}$ for the quadratic Blaschke product

$$B(z) = \frac{3z^2 + 1}{3 + z^2} \tag{2.31}$$

We prove:

Theorem 2.9 *Let f be an analytic function with a normalized simple parabolic point at the origin. Assume that the immediate basin B_0^f is simply-connected, and $f : B_0^f \to B_0^f$ is a degree-2 branched covering map. Then there is a conformal isomorphism $\varphi : B_0^f \to \mathbb{D}$ so that*

$$f = \varphi^{-1} \circ B \circ \varphi \quad on \ B_0^f,$$

where $B(z)$ is as in (2.31). In particular, any two analytic maps satisfying the conditions of the theorem are conformally conjugate on their respective B_0's.

Proof For each sufficiently large $j \in \mathbb{N}$, let σ_j denote that component of the intersection with B_0^f of the circle of radius $1/j$ about 0 which contains $-1/j$. Thus for large j, the arc σ_j is almost the whole circle: a small closed arc near the positive real axis has been cut out. It is immediate that each σ_j is a crosscut of B_0^f and that σ_j's for different j's are disjoint (see Fig. 2.3a). Let N_j be the crosscut neighborhood of σ_j that does not intersect σ_{j-1}. It is evident that $\sigma_{j+1} \subset N_j$. The diameter of σ_j is at most $2/j \to 0$, hence the crosscut neighborhoods N_j form a fundamental chain. It is clear that the impression of this fundamental chain contains 0. Recalling that a prime end is an equivalence class of fundamental chains, we denote by $\hat{0}$ the prime end containing the above fundamental chain.

(a)   **(b)**

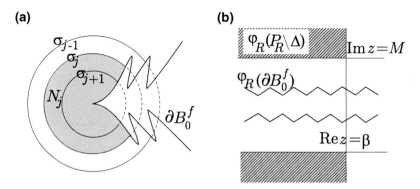

Fig. 2.3 An illustration of the proof of Theorem 2.9

By the Riemann Mapping Theorem, there is a conformal isomorphism φ from B_0^f to $\mathbb{D}$, mapping the unique critical point of f in B_0^f to 0. By Carathéodory theory, φ extends to map the set of prime ends of B_0^f homeomorphically to the unit circle. We can further require that the extension of φ send $\widehat{0}$ to 1; with this additional condition, φ is unique.

We use $z_n \underset{\text{p.e.}}{\to} \widehat{0}$ as an abbreviation for the assertion that z_n is eventually in N_j for each j. From the construction of the topology on the space of prime ends and of the Carathéodory extension, we extract the following:

Continuity at 0 *Let (z_n) be a sequence in B_0^f. Then $\varphi(z_n) \to 1$ if and only if $z_n \underset{\text{p.e.}}{\to} \widehat{0}$.*
In view of the construction of the crosscuts σ_j, it might appear at first glance that the requirement that $z_n \underset{\text{p.e.}}{\to} \widehat{0}$ be more or less the same as the requirement that $z_n \to 0$.

However, proving this requires more control over the structure of ∂B_0^f than we have. We next develop a convenient condition which suffices to guarantee convergence of $\varphi(z_n)$ to 1.

Let P_R be a repelling petal which maps under ϕ_R to a left half-plane

$$\{u + iv : u < \beta\}.$$

To give some room for maneuver, we assume that there is a larger petal P_R' mapping to the half-plane $\{u < \beta'\}$ for some $\beta' > \beta$. For the next few paragraphs, ϕ_R^{-1} will denote the inverse of the restriction of ϕ_R to P_R'.

We next need to argue that $\text{Im}(\phi_R)$ is bounded on $P_R \setminus B_0^f$. More precisely, let M be a positive number such that if $z \in P_R$ with

$$|\text{Im}(\phi_R(z))| \geq M,$$

then $z \in B_0^f$ (see Fig. 2.3b).

Now let $\Delta := \{z \in P_R : |\operatorname{Im}(\phi_R(z))| \leq M\}$. □

Lemma 2.7 *If (z_n) is a sequence in $B_0^f \setminus \Delta$ such that $z_n \to 0$, then $z_n \xrightarrow[p.e]{} \widehat{0}$.*

Proof (Lemma 2.7) Recall the definition of the crosscut neighborhoods N_j above, and let D_j denote the open disk of radius $1/j$ about 0. Let γ_1 and γ_2 denote the segments of the upper and lower edges of Δ which start at 0 and extend to the first point where the edges in question meet the bounding circle of D_j.

Then γ_1 and γ_2 are crosscuts of N_j; deleting them splits N_j into 3 subdomains, one of which is bounded by γ_1, γ_2, and a left-facing arc of the circle bounding D_j. All we want to extract from the above is that this latter domain is contained in N_j. On the other hand, elementary topological considerations, resting on the fact that the two edges of Δ are smooth arcs tangent to the positive imaginary axis at 0, show that the domain described above is $D_j \setminus \Delta$ for large enough j. In short,

$$D_j \setminus \Delta \subset N_j \quad \text{for sufficiently large } j.$$

Now fix j large enough so that the above holds. If z_n converges to 0 staying outside of Δ, it is eventually in $D_j \setminus \Delta$, hence, eventually in N_j. This holds for all sufficiently large j, so $z_n \xrightarrow[p.e]{} \widehat{0}$, as asserted.

We return to the proof of Theorem 2.9. By assumption, $f : B_0^f \to B_0^f$ is a degree-2 branched covering, and B is the conjugate of f on B_0^f under a conformal isomorphism $\varphi : B_0^f \to \mathbb{D}$. Thus, B is a degree-2 branched covering of $\mathbb{D}$ to itself. It is a standard fact, following from an easy application of the Schwarz Lemma, that such a B has the form

$$B(z) = e^{i\theta} \frac{z^2 - a}{1 - \overline{a}z^2},$$

where $a \in \mathbb{D}$ and $\theta \in \mathbb{R}$. By this formula, B extends analytically to a neighborhood of the closed disk. To complete the proof, we just need to determine θ and a.

For any $z_0 \in B_0^f$, $f^n(z_0)$ converges to 0 from the negative real direction, so z_n is eventually outside Δ, and, by Lemma 2.7 $z_n \xrightarrow[p.e]{} \widehat{0}$, so

$$\varphi(f^n(z_0)) = B^n(\varphi(z_0)) \to 1,$$

from which it follows that 1 is a fixed point of B:

$$B(1) = 1.$$

Since B maps the unit circle to itself, preserving orientation, $B'(1)$ is real and positive. Since the orbit $B^n(\varphi(z_0)) \to 1$, it is not possible that $B'(1) > 1$.

Now specialize to $z_0 \in B_0^f \cap P_R$ with $\mathrm{Im}(\phi_R(z_0)) > M$. Let f^{-1} denote the inverse of the restriction of f to P_R. Then, for any $n \geq 0$, $f^{-n}(z_0) \in P_R$ and $\mathrm{Im}(\phi_R(f^{-n}(z_0)) = \mathrm{Im}(\phi_R(z_0)) > M$, so $f^{-n}(z_0) \in B_0^f \setminus \Delta$. Since z_0 is in the repelling petal P_R, $f^{-n}(z_0) \to 0$, so, by Lemma 2.7, $y_{-n} := \varphi(f^{-n}(z_0)) \to 1$. The sequence (y_{-n}) is a backward orbit for B: $B(y_{-n-1}) = y_{-n}$. The existence of just one backward B-orbit converging to the fixed point 1 of B implies

- $B'(1)$ cannot be < 1, so $B'(1)$ must be equal to 1,
- $B''(1)$ cannot be non-zero; if it were, any backward orbit converging to 1 would have to do so from the positive real direction, in particular, from outside $\mathbb{D}$, whereas our sequence y_{-n} is inside $\mathbb{D}$.

We thus have the general algebraic formula for $B(z)$ and the conditions:

$$B(1) = 1, \quad B'(1) = 1, \quad B''(1) = 0;$$

routine algebra then shows that $B(z)$ must be as in the statement of the theorem. $\square$

Chapter 3
Global Theory

Abstract We discuss the covering properties of the parabolic renormalization of the quadratic map and describe a class of maps with global covering properties and Jordan basins, invariant under the parabolic renormalization operator.

Keywords Branched covering · Map of finite type · Koebe function · Parabolic basin · Renormalization-invariant class

3.1 Basic Facts About Branched Coverings

We give a brief summary of relevant facts about analytic branched coverings. A more detailed exposition is found e.g., in Appendix E of [Mil1].

For a holomorphic map $f : X \to Y$ between two Riemann surfaces, a *regular value* is a point $y \in Y$ for which one can find an open neighborhood $U = U(y)$ such that

$$f^{-1}(U) \xrightarrow{f} U$$

is a covering map. The complement of this set consists of the *singular values* of f, and will be denoted $\mathrm{Sing}(f)$.

By definition, $y \in Y$ is an *asymptotic value* of f if there exists a parametrized path

$$\gamma : (0, 1) \to X, \quad \text{such that} \quad \lim_{t \to 1-} f(\gamma(t)) = y,$$

and such that $\lim_{t \to 1-} \gamma(t)$ does not exist in X. We will be concerned with the situation when X is a proper subdomain of the Riemann sphere. In this case, the non-existence of the limit can be replaced with

$$\gamma(t) \xrightarrow[t \to 1-]{} \partial X.$$

We will denote $\mathrm{Asym}(f) \subset Y$ the set of all asymptotic values of f.

O.E. Lanford III and M. Yampolsky, *Fixed Point of the Parabolic Renormalization Operator*, SpringerBriefs in Mathematics, DOI 10.1007/978-3-319-11707-2_3

Recall that $y_0 \in Y$ is called a *critical value* (or a *ramified point*) of f if there exists $x_0 \in X$ such that $y_0 = f(x_0)$, and the local degree of f at x_0 is $n \geq 2$. Thus, in local coordinates, one has

$$f(x) - y_0 = c(x - x_0)^n + \cdots \text{ where } n \geq 2.$$

The point x_0 is called a *critical point* of f (or a *ramification point*). We denote $\mathrm{Crit}(f) \subset X$ the set of the critical points of f. One then has:

Proposition 3.1 *For a holomorphic map $f : X \to Y$ between Riemann surfaces,*

$$\mathrm{Sing}(f) = \overline{\mathrm{Asym}(f) \cup f(\mathrm{Crit}(f))};$$

that is, the set of the singular values of f is the closure of the union of its asymptotic and critical values.

Recall that a nonconstant map $g : S_1 \to S_2$ between two Hausdorff topological spaces is called *proper* if the preimage of every compact set of S_2 is compact in S_1. It is easy to see the following.

Proposition 3.2 *Let $f : X \to Y$ be a proper holomorphic map of Riemann surfaces. Let W be a nonempty connected open subset of Y, and let V be a connected component of $f^{-1}(W)$. Then $f : V \to W$ is also proper.*

For a holomorphic map $f : X \to Y$, define the degree of f at $y \in Y$ (denoted $\deg_y(f)$) as the possibly infinite sum of the number of preimages of y in X, counted with multiplicity. It is not difficult to see that a proper analytic map has a well-defined local degree:

Proposition 3.3 *If $f : X \to Y$ is a proper analytic map between Riemann surfaces, and Y is connected, then $\deg_y(f)$ is finite and independent of y (and can be denoted as $\deg(f)$).*

As a consequence, note the following.

Proposition 3.4 *If $f : X \to Y$ is a nonconstant proper analytic map between Riemann surfaces, and Y is connected, then $f(X) = Y$.*

Proposition 3.5 *Let $g : S_1 \to S_2$ be a proper continuous map between Hausdorff topological spaces which is everywhere a local homeomorphism. Then f is a covering map.*

In particular, putting together Proposition 3.3 and Proposition 3.5, we have

Proposition 3.6 *Let $f : X \to Y$ be a proper analytic mapping between Riemann surfaces. Let Y be connected, and set $d = \deg(f)$. Then*

$$f : X \backslash \mathrm{Crit}(f) \to Y \backslash f(\mathrm{Crit}(f))$$

is a degree d covering.

In view of the above, a proper analytic map is sometimes called a *branched covering of a finite degree*. Generalizing to the case when the local degree is infinite gives the following definition:

Definition 3.1 A holomorphic map $f : X \to Y$ between Riemann surfaces is a *branched covering* if every point $y \in Y$ has a connected neighborhood $U \equiv U(y)$ such that the restriction of f to each connected component of $f^{-1}(U)$ is a proper map.

As a canonical example, consider the case $X = Y = \hat{\mathbb{C}}$. Nonconstant rational maps f are clearly branched coverings; and every branched covering is, in fact, a rational map.

Let us formulate another general lemma which we will find useful:

Lemma 3.1 *Let $f : U \to V$ be a proper analytic map between connected subdomains of $\hat{\mathbb{C}}$. Assume that f has a single critical value $v \in V$, and that V is a topological disk. Then U is also a topological disk, and f has only one critical point u in U, such that $f^{-1}(v) = \{u\}$.*

Proof Set $\hat{V} \equiv V \setminus \{v\}$ and $\hat{U} \equiv f^{-1}(\hat{V})$. By Proposition 3.6, the map $f : \hat{U} \to \hat{V}$ is a covering. The domain $\hat{V}$ is homeomorphic to $\mathbb{D} \setminus \{0\}$. By standard facts about coverings of $\mathbb{D} \setminus \{0\}$, the domain $\hat{U}$ is also homeomorphic to the punctured disk. Moreover, denoting $\eta_{\hat{V}}$ and $\eta_{\hat{U}}$ conformal homeomorphisms mapping $\mathbb{D} \setminus \{0\}$ to $\hat{V}$ and $\hat{U}$ respectively, we see that

$$\eta_{\hat{V}}(f(z)) = \text{const} \cdot (\eta_{\hat{U}}(z))^d \text{ for } d \geq 2. \qquad \square$$

3.2 Parabolic Renormalization of the Quadratic Map

A key example for our investigation is the quadratic polynomial

$$f_0(z) = z \mapsto z + z^2.$$

It is well known that

- The basin of attraction $B^{f_0} =: B_0$ of the parabolic point is connected.
- $f : B_0 \to B_0$ is a branched covering of degree 2.

We will prove more detailed or more general versions of these assertions later; see, in particular, the proof of Theorem 3.2, so we will not repeat the proofs here.

By a trivial computation, $-1/2$ is a critical point of f_0; it is the only one in the finite plane, and the corresponding critical value is $-1/4$. Let ϕ_A be an attracting Fatou coordinate for f_0, defined on all of B_0. Using Proposition 2.12 and taking into account the surjectivity of f_0 on B_0, we get

Proposition 3.7 *The critical points of ϕ_A are the preimages, under iterates of f_0, of the critical value $-1/4$:*

$$\mathrm{Crit}(\phi_A) = \cup_{n \geq 1} f_0^{-n}(-1/4).$$

All critical points have local degree 2, and their images are integer translates of each other:

$$\phi_A(\mathrm{Crit}(\phi_A)) = \{\phi_A(-1/4) - n, n \geq 1\}.$$

By standard results of complex dynamics,

$$\overline{\mathrm{Crit}(\phi_A)} = \mathrm{Crit}(\phi_A) \cup J(f_0). \qquad (3.1)$$

In fact (cf. [Do2]), ϕ_A, mapping B_0 to $\mathbb{C}$, is a branched covering (but we will not need this result).

Our next objective is the following:

Theorem 3.1 *Let (h^+, h^-) be an Écalle-Voronin pair for f_0 (i.e., one constructed with some choice of normalization for ϕ_A and ϕ_R). Then the germs h^+ (at 0) and h^- (at ∞) have maximal analytic continuations to Jordan domains $\hat{W}^+$ and $\hat{W}^-$ (in the sphere $\hat{\mathbb{C}}$).*

See Fig. 4.1. Theorem 3.1 follows from Theorem 2.8 and a well-known folklore result:

Theorem 3.2 *The Julia set of the quadratic polynomial $f_0(z) = z + z^2$ is a Jordan curve. Orbits starting inside this curve converge to 0; those starting outside converge to ∞. The dynamics of f_0 restricted to the Julia set is topologically conjugate to the angle-doubling map of the circle; specifically, there exists a unique continuous map $\rho : J(f_0) \to \mathbb{T}$ such that*

$$\rho(f_0(z)) = 2\rho(z) \bmod 1.$$

In the next subsection, we will provide a proof of Theorem 3.2; while it is fairly standard, the strategy used will be useful to us later in a more complicated situation.

3.2.1 The Julia Set of $f_0(z) = z + z^2$ Is a Jordan Curve

We begin by defining several useful local inverse branches of f_0 by choosing the appropriate branches of the square root in the formal expression

$$f_0^{-1}(w) = \sqrt{w + 1/4} - 1/2. \qquad (3.2)$$

We first cut the plane along the positive real ray $R \equiv [-1/4, +\infty)$ and denote

$$g_0 : \mathbb{C}\backslash R \to \mathbb{H}$$

the inverse branch with nonnegative imaginary part. Similarly, we define

$$g_1 : \mathbb{C}\backslash R \to -\mathbb{H}$$

to be the inverse branch with negative imaginary part, so that $g_1 \equiv 1 - g_0$.

Finally, we define an inverse branch g which fixes the parabolic point $z = 0$. We slit the plane along the ray $(-\infty, -1/4]$, and select the branch of the square root in (3.2) with nonnegative real part. In this way, we get

$$g : \mathbb{C}\backslash(-\infty, -1/4] \to \{\mathrm{Re}\, z > -1/2\}.$$

We note that Taylor expansion of g at $w = 0$ begins with $w - w^2 + \cdots$, and thus it also has a parabolic fixed point at the origin.

When needed, we will continuously extend the three inverse branches defined above to the edges of the slits.

Proposition 3.8 *The branch g maps $\mathbb{C}\backslash(-\infty, 0]$ into itself.*

Proof As already noted, the image of g is contained in the half-plane

$$\{\mathrm{Re}\, z > -1/2\}.$$

Furthermore, since g is a branch of the inverse of f_0, if $g(w) \in [-1/2, 0)$, then $w \in f_0([-1/2, 0)) = ([-1/4, 0) \subset (-\infty, 0)$, so the image under g of $\mathbb{C}\backslash(-\infty, 0]$ does not intersect $[-1/2, 0)$, and thus is contained in $\mathbb{C}\backslash(-\infty, 0]$.

Applying the Denjoy-Wolff Theorem, we obtain:

Corollary 3.1 *The successive iterates g^n converge to the constant 0, the parabolic point, uniformly on compact subsets of $\mathbb{C}\backslash(-\infty, 0]$.*

Next, we observe:

Proposition 3.9 *The intersection*

$$J(f_0) \cap \mathbb{R} = \{-1, 0\}.$$

Proof Indeed, if $x > 0$, then

$$f(x) = x + x^2 > x, \text{ and, furthermore, } f^n(x) > x + nx^2 \to \infty,$$

and hence $(0, \infty) \subset F(f_0)$. Since

$$f_0((-\infty, -1)) = (0, \infty),$$

the same holds for $(-\infty, -1)$. Finally, f_0 maps the interval $(-1, 0)$ to itself and $f_0(x) > x$ for x in this interval, so $f_0^n(x) \to 0$. Hence, $(-1, 0) \subset F(f)$ as well, as claimed.

Define

$$\mathbf{A} := \{z : 1/2 < |z + 1/2| < 2\}.$$

i.e., $\mathbf{A}$ is an annulus, centered at the critical point $-1/2$, with inner radius $1/2$ and outer radius 2. We collect some elementary facts about this annulus into the following proposition:

Proposition 3.10

1. *If $|z + 1/2| \geq 2$—i.e., if z is outside the annulus, then $|f_0(z) + 1/2| > |z + 1/2|$, so $f_0(z)$ is again outside the annulus, and $|f_0^n(z) + 1/2|$ goes monotonically to ∞.*
2. *If $|z+1/2| < 1/2$—i.e., if z is strictly inside the annulus, then $|f_0(z)+1/4| < 1/4$.*
3. *If $|z + 1/2| < 1/2$, then $f_0^n(z) \to 0$.*
4. *If $f(z) \in \mathbf{A}$, then $z \in \mathbf{A}$, i.e., $f_0^{-1}\mathbf{A} \subset \mathbf{A}$.*
5. *$J(f_0)$ is contained in the closure $\overline{\mathbf{A}}$ of $\mathbf{A}$.*

Proof Items (1) and (2) follow from elementary estimates. For (3), since the open disk of radius $1/4$ about $-1/4$ is contained in the open disk of radius $1/2$ about $-1/2$, this latter disk is mapped into itself by f_0 and hence by all its iterates f_0^n. In particular, this sequence of iterates is uniformly bounded on the disk in question and so, by Montel's Theorem, is a normal family. Furthermore, elementary considerations show that for $-1 < x < 0$, $f_0^n(x) \to 0$. It follows then from Vitali's Theorem that f_0^n converges to the constant 0 uniformly on compact sets of the open disk $\{|z + 1/2| < 1/2\}$.

From (1) and (2), if $z \notin \mathbf{A}$, then $f_0(z) \notin \mathbf{A}$, which is tautologically equivalent to (4). Finally, from (1), if z is outside $\overline{\mathbf{A}}$, then $f_0^n(z) \to \infty$, so $z \notin J(f_0)$, and, from (3), if z is inside $\overline{\mathbf{A}}$, $f_0^n(z) \to 0$, and so, again, $z \notin J(f_0)$.

To prove Theorem 3.2, we are going to show that $f_0^{-n}\mathbf{A}$ are a decreasing sequence of topological annuli which shrink down to a Jordan curve J. Points inside J are attracted to the parabolic point; those outside are attracted to ∞. Thus, the Jordan curve J contains the Julia set. Our argument also shows that the preimages (of arbitrary order) of the parabolic point 0 under f_0 are dense in J, so the Julia set is all of J.

We define first

$$\mathbf{A}_0 := \{z \in \mathbf{A} : \text{Im}\, z > 0\} \quad \text{and} \quad \mathbf{A}_1 := \{z \in \mathbf{A} : \text{Im}\, z < 0\},$$

and then, for any $n = 2, \ldots$ and any sequence $i_0 i_1 \ldots i_{n-1}$ of n 0's and 1's,

$$\mathbf{A}_{i_0 i_1 \ldots i_{n-1}} := \{z \in \mathbf{A}_{i_0} : f_0^j(z) \in \mathbf{A}_{i_j} \text{ for } 1 \leq j < n\}.$$

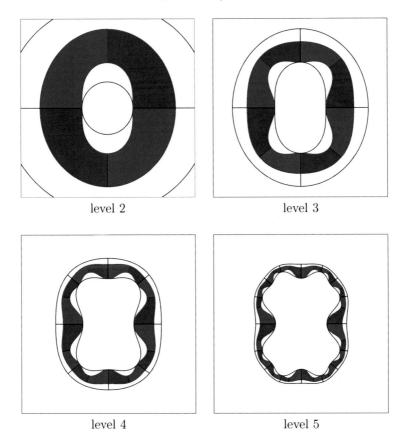

level 2   level 3

level 4   level 5

Fig. 3.1 There are 2^n puzzle pieces of level n. In the picture for level n, we also show the outlines of the puzzle pieces of level $n - 1$, each of which contains two pieces of level n. The pieces of level n can be labeled (successively, counterclockwise, starting from the positive real axis) with $0, 1, \ldots, 2^n - 1$; the labeling used in the text is the binary representation of this one

We will refer to the sets $\mathbf{A}_{i_0 \ldots i_{n-1}}$ as *puzzle pieces* (*of level* n) (Fig. 3.1). It is easy to see that

- $\mathbf{A}_{i_0 i_1 \ldots i_{n-1}}$ is decreasing in n: $\mathbf{A}_{i_0 \ldots i_{n-1}} \supset \mathbf{A}_{i_0 \ldots i_n}$;
- $f \mathbf{A}_{i_0 i_1 \ldots i_n} = \mathbf{A}_{i_1 \ldots i_n}$;
- $\mathbf{A}_{i_0 i_1 \ldots i_n} = g_{i_0} \mathbf{A}_{i_1 \ldots i_n}$, and hence
  $\mathbf{A}_{i_0 i_1 \ldots i_{n-1}} = g_{i_0} \circ g_{i_1} \circ \cdots \circ g_{i_{j-1}} \mathbf{A}_{i_j i_{j+1} \ldots i_{n-1}}$ for $1 \leq j < n - 1$.

The main step in the argument is to show that the diameters of the puzzle pieces $\mathbf{A}_{i_0 i_1 \ldots i_{n-1}}$ go to zero as $n \to \infty$, uniformly in $i_0 i_1 \ldots i_{n-1}$. The strategy that we use is standard; it also appears in, e.g., [Lyu], Proposition 1.10.

Concretely, we define

$$\rho_n := \sup \left\{ \operatorname{diam}(\mathbf{A}_{i_0 \ldots i_{n-1}}) : i_0 \ldots i_{n-1} \in \{0, 1\}^n \right\}.$$

Since $\mathbf{A}_{i_0...i_{n-1}} \supset \mathbf{A}_{i_0...i_n}$, the sequence ρ_n is non-increasing in n, so

$$\rho_* := \lim_{n\to\infty} \rho_n \tag{3.3}$$

exists.

We will prove:

Proposition 3.11 *The limit* $\rho_* = 0$.

The first step is to argue that it suffices to consider binary sequences one at a time:

Lemma 3.2 *There is an infinite sequence* $i_0 i_1 \ldots i_n \ldots$ *so that*

$$\mathrm{diam}(A_{i_0 i_1 ... i_n}) \geq \rho_* \quad \textit{for all } n,$$

i.e., a single *descending chain of puzzle pieces with diameter converging to* ρ_*.

Proof (Lemma 3.2) A simple diagonal argument gives the existence of a sequence $i_0 i_1 \ldots i_k \ldots$ such that for all k,

$$\lim_{n\to\infty} \sup \left\{ \mathrm{diam}(\mathbf{A}_{i_0,...i_{k-1}i'_k...i'_{k+n-1}}) : i'_k \ldots i'_{k+n-1} \in \{0,1\}^n \right\} \geq \rho_*$$

Since $\mathrm{diam}(\mathbf{A}_{i_0 i_1 ... i_{n-1}})$ is non-increasing in n, it follows that

$$\mathrm{diam}(\mathbf{A}_{i_0 i_1 ... i_k}) \geq \rho_* \quad \text{for all } k,$$

proving the lemma.

Proof (Proposition 3.11) We assume the contrary, that $\rho_* > 0$ and we fix a sequence $i_0 i_1 \ldots$ as in the Lemma 3.2, i.e., so that $\mathrm{diam}(\mathbf{A}_{i_0 i_1 ... i_{k-1}})$ does not go to zero as $k \to \infty$. We consider separately the three cases:

1. i_j is eventually 0;
2. i_j is eventually 1;
3. i_j takes both values 0 and 1 infinitely often.

We start with case (3). There are then infinitely many j's so that

$$i_j = 0 \quad \text{and} \quad i_{j+1} = 1.$$

Let j_k be a strictly increasing sequence of such j's. Then, for each k,

$$f_{-j_k} := g_{i_0} \circ g_{i_1} \circ \cdots \circ g_{i_{j_k-1}}$$

is an analytic branch of the inverse of $f_0^{j_k}$ mapping $\mathbf{A}_{01}$ bijectively to $\mathbf{A}_{i_0 ... i_{j_k+1}}$. The closure of $\mathbf{A}_{01}$ does not intersect the closure of the postcritical set of f_0; we let U be a bounded simply connected open neighborhood of $\overline{\mathbf{A}_{01}}$, disjoint from the postcritical set. Then each f_{-j_k} has an extension to a branch of the inverse of $f_0^{j_k}$, defined and analytic on U; we denote this extension also by f_{-j_k}.

We next want to argue that the f_{-j_k} are uniformly bounded on U and hence form a normal family there. We can see this as follows: Let $\hat{U}$ be a neighborhood of ∞ which is forward-invariant under f_0 and disjoint from U. Then, if $f_{j_k}(z)$ is in $\hat{U}$ (for some k and some $z \in U$), we will have, since f_{j_k} is an inverse branch of $f_0^{j_k}$ and $\hat{U}$ is forward invariant,

$$z = f_0^{j_k}(f_{-j_k}(z)) \in \hat{U},$$

which is impossible since by hypothesis $z \in U$ and U is disjoint from $\hat{U}$. Thus, the sets $f_{-j_k}U$ are all disjoint from $\hat{U}$, so (f_{-j_k}) is uniformly bounded on U, as desired.

By Montel's Theorem, there is a subsequence of (j_k) along which f_{-j_k} converges uniformly on compact subsets of U. By adjusting the notation, we can assume that the sequence (f_{-j_k}) itself converges; we denote its limit by h. Since

$$\mathrm{diam}(f_{-j_k}\mathbf{A}_{01}) = \mathrm{diam}(\mathbf{A}_{i_0 \ldots i_{j_k+1}})$$

does not go to zero as $k \to \infty$, and since f_{-j_k} converges uniformly on the closure of $\mathbf{A}_{01}$, the limit h is nonconstant.

Let $z_0 := g_0(g_1(-1)) \in \mathbf{A}_{01} \cap J(f_0)$. Since the Julia set if backward-invariant under f_0 and closed, then

$$w_0 := h(z_0) = \lim_{k \to \infty} f_{-j_k}(z_0)$$

is again in the Julia set. Since h is nonconstant, $h\mathbf{A}_{01}$ is an open neighborhood of w_0; let W be another open neighborhood whose closure is compact and contained in $h\mathbf{A}_{01}$. A straightforward application of Rouché's theorem shows that

$$f_{-j_k}\mathbf{A}_{01} \supset W \quad \text{for sufficiently large } k.$$

Since f_{-j_k} is a branch of the inverse of $f_0^{j_k}$, it follows that

$$f_0^{j_k}W \subset \mathbf{A}_{01} \quad \text{for all sufficiently large } k. \tag{3.4}$$

We are going to deduce from (3.4) that

$$|f_0^n(w) + 1/2| \leq 2 \quad \text{for } w \in W \text{ and } n \geq 0,$$

which by Montel's theorem contradicts the fact that $w_0 \in W$ is in the Julia set. Suppose therefore that $|f_0^n(w) + 1/2| > 2$ for some $w \in W$ and some n. Then, by Proposition 3.10 (1),

$$|f_0^{j_k}(w) + 1/2| > 2 \quad \text{for all } k \text{ with } j_k \geq n.$$

Since

$$|z + 1/2| \leq 2 \quad \text{for all } z \in \mathbf{A},$$

this contradicts (3.4), and so completes the proof in case 3.

We turn next to case (1) above, and deal first with the situation $i_j = 0$ for all j. We write temporarily

$$\mathbf{A}^{(n)} := \mathbf{A}_{\underbrace{0\ldots0}_{n \text{ terms}}} = (g_0)^{n-2} \mathbf{A}_{00}.$$

Now g_0 maps $\mathbf{A}_{00}$ into itself, and $g_0 = g$ on $\mathbf{A}_{00}$, so we can write

$$\mathbf{A}^{(n)} = g^{n-2} \mathbf{A}_{00}.$$

By Corollary 3.1 and local properties of parabolic dynamics near 0,

$$g^{n-2} \to 0 \quad \text{uniformly on } \mathbf{A}_{00},$$

so

$$\mathrm{diam}(\mathbf{A}^{(n)}) \to 0 \quad \text{as } n \to \infty.$$

Next consider sequences of the form $i_0 i_1 \ldots i_k 00 \ldots$, and use the formula

$$\mathbf{A}_{i_0 \ldots i_k \underbrace{0 \cdots 0}_{n \text{ terms}}} = g_{i_0} \circ g_{i_1} \circ \cdots \circ g_{i_k} A^{(n)}.$$

The mapping $g_{i_0} \circ g_{i_1} \circ \cdots \circ g_{i_k}$ extends to be continuous on $\overline{\mathbf{A}_{00}}$, and $\mathrm{diam}(A^{(n)}) \to 0$ by what we just proved, so

$$\mathrm{diam}(\mathbf{A}_{i_0 \ldots i_k \underbrace{0 \cdots 0}_{n \text{ terms}}}) \to 0 \quad \text{as } n \to \infty.$$

A similar argument, using $g_1 = g$ on $\mathbf{A}_{11}$ shows that

$$\mathrm{diam}(\mathbf{A}_{i_0 \ldots i_k \underbrace{1 \cdots 1}_{n \text{ terms}}}) \to 0 \quad \text{as } n \to \infty.$$

Let $\mathbf{i} = (i_j)_{j=0}^{N}$ be a finite or infinite sequence ($N \leq \infty$) of 0's and 1's. We interpret it as a binary representation of an element of the circle $\mathbb{R}/\mathbb{Z}$:

$$\underline{i_2} \equiv \sum_{j=0}^{N} i_j 2^{j+1} \bmod \mathbb{Z}.$$

For any such dyadic sequence,

$$\overline{\mathbf{A}_{i_0}} \supset \overline{\mathbf{A}_{i_0 i_1}} \supset \cdots \supset \overline{\mathbf{A}_{i_0 \ldots i_n}} \supset \cdots$$

is a nested sequence of compact sets in $\mathbb{C}$ with diameter going to 0, so its intersection contains exactly one point, which we denote by $\hat{z}(\mathbf{i})$. Since any $\mathbf{A}_{i_0 \ldots i_{n-1}}$ contains $g_{i_0} g_{i_1} \ldots g_{i_{n-1}}(-1) \in J(f_0)$, each $\hat{z}(\mathbf{i}) \in J(f_0)$. It is immediate from the construction that $\mathbf{i} \mapsto \hat{z}(\mathbf{i})$ is continuous from $\{0, 1\}^{\mathbb{N}}$ to $\mathbb{C}$. It is not injective, but its non-injectivity is of a simple and familiar nature:

Lemma 3.3 *Let* $\mathbf{i} = i_0 i_1 \ldots i_{n-1}$ *and* $\mathbf{i}' = i'_0 i'_1 \ldots i'_{n-1}$ *be two finite dyadic sequences of equal length such that*

$$\overline{\mathbf{A}_{\mathbf{i}}} \cap \overline{\mathbf{A}_{\mathbf{i}'}} \neq \emptyset.$$

Then either $\underline{i_2} = \underline{i}'_2$ *or* $\underline{i_2} = \underline{i}'_2 \pm 2^{-n}$.

The proof is a straightforward induction in n which we leave to the reader. As a corollary, we get:

Corollary 3.2 *Let* $\mathbf{i}$ *and* $\mathbf{i}'$ *be two infinite dyadic sequences. Then*

$$\hat{z}(\mathbf{i}) = \hat{z}(\mathbf{i}') \quad \text{if and only if} \quad \underline{i_2} = \underline{i}'_2.$$

Proof For any $\mathbf{i} = i_0 \ldots i_n \ldots$, and any n

$$|\underline{i_2} - \underline{i_0 \ldots i_{n-1}}_2| \leq 2^{-n}.$$

Hence, if $\underline{i_2} \neq \underline{i}'_2$ and n is large enough so that

$$|\underline{i_2} - \underline{i}'_2| > 3 \times 2^{-n},$$

then

$$|\underline{i_0 \ldots i_{n-1}}_2 - \underline{i'_0 \ldots i'_{n-1}}_2| > 2^{-n},$$

which by Lemma 3.3 implies that $\overline{\mathbf{A}_{i_0 \cdots i_{n-1}}}$ and $\overline{\mathbf{A}_{i'_0 \cdots i'_{n-1}}}$ are disjoint, and hence that $\hat{z}(\mathbf{i}) \neq \hat{z}(\mathbf{i}')$.

Conversely, if $\underline{i_2} = \underline{i}'_2$, then either $\mathbf{i} = \mathbf{i}'$ or, possibly after interchanging the two sequences, $\mathbf{i}$ has the form $i_0 i_1 \ldots i_{m-1} 0111 \ldots$ and $\mathbf{i}'$ the form $i_0 i_1 \ldots i_{m-1} 1000 \ldots$, and from this it follows easily that $\mathbf{A}_{i_0 \cdots i_n}$ and $\mathbf{A}_{i'_0 \cdots i'_n}$ intersect for all n and hence that $\hat{z}(\mathbf{i}) = \hat{z}(\mathbf{i}')$.

Hence, there is an injective mapping $\theta \mapsto z(\theta)$, from the circle $\mathbb{R}/\mathbb{Z}$ to $\mathbb{C}$, such that

$$\hat{z}(\mathbf{i}) = z(\mathbf{i}_2) \quad \text{for all } \mathbf{i}.$$

Proposition 3.12 *The mapping $z(.)$ is a homeomorphism from the circle onto the image J of $\mathbf{i} \mapsto \hat{z}(\mathbf{i})$. In particular, J is a Jordan curve.*

Proof $z(.)$ is injective by construction, and its continuity follows easily from the fact that $\operatorname{diam}(\mathbf{A}_{i_0 \cdots i_{n-1}}) \to 0$. It is therefore a homeomorphism by the standard topological fact that an injective continuous mapping from a compact space (to a Hausdorff space) is a homeomorphism.

It remains to show

Proposition 3.13 $J = J(f_0) = \bigcap_{n \geq 0} f_0^{-n} \overline{A}$.

Proof We show the following.

1. $J \subset J(f_0)$,
2. $J(f_0) \subset \bigcap_{n \geq 0} f_0^{-n} \overline{A}$,
3. $\bigcap_{n \geq 0} f_0^{-n} \overline{A} \subset J$.

(1) If $z \in J$, then $\{z\} = \bigcap \overline{A_{i_0 \cdots i_{n-1}}}$ for some sequence $i_0 i_1 \ldots$. Since $A_{i_0 \cdots i_{n-1}}$ contains $g_{i_0} \circ \cdots \circ g_{i_{n-1}}(-1) \in J(f_0)$, and since $\operatorname{diam}(A_{i_0 \cdots i_{n-1}}) \to 0$, $z \in \overline{J(f_0)} = J(f_0)$.

(2) This proof has essentially already been given: If $z \notin \bigcap_{n \geq 0} f_0^{-n} \overline{A}$, then some $f_0^n(z) \notin \overline{A}$, so, by the elementary properties of the basic annulus $\mathbf{A}$ (Proposition 3.10), $f_0^n(z)$ converges either to 0 or to ∞, so $z \notin J(f_0)$.

(3) Assume $z \in \bigcap_{n \geq 0} f_0^{-n} \overline{A}$, i.e., $f_0^n(z) \in \overline{A}$ for all n. We consider two cases:

- $f_0^n(z) \notin \mathbb{R}$ for all n. Put $i_n = 0$ of 1 according to whether $f_0^n(z)$ is in the upper or lower half-plane. Then $z \in \mathbf{A}_{i_0 \cdots i_{n-1}}$ for all n, so $z = \hat{z}(\mathbf{i}) \in J$.
- $f_0^n(z) \in \mathbb{R}$ for some n. The intersection of $\overline{A}$ with $\mathbb{R}$ is the union of the two intervals $[-2.5, -1]$ and $[0, 1.5]$, and the first of these maps into the positive real axis, which is mapped into itself. Hence $f_0^n(z)$ is eventually in $[0, 1.5]$. But the only orbit staying forever in $[0, 1.5]$ is the fixed point 0, so z must be a preimage of finite order of 0. It is then straightforward to show that there is a sequence of the form $i_0 i_1 \ldots i_{n-1} 000 \ldots$ with $z = \hat{z}(\mathbf{i})$.

This completes the proof of Theorem 3.2. We note further that

$$f_0(\hat{z}(i_0 i_1 i_2 \ldots)) = \hat{z}(i_1 i_2 \ldots) \quad \text{for all binary sequences } i_0 i_1 \ldots,$$

from which it follows that we have

Proposition 3.14 *The map $\theta \mapsto z(\theta)$ conjugates $\theta \mapsto 2 \cdot \theta$ on $\mathbb{R}/\mathbb{Z}$ to f_0 on its Julia set $J(f_0)$.*

3.3 Covering Properties of the Écalle-Voronin Invariant of f_0

For a parabolic germ $f(z)$ of the form (2.2) we will denote $\mathbf{h}_f$ its Écalle-Voronin invariant. We remind the reader that the pair of germs $\mathbf{h}_f = (h^+, h^-)$ is defined up to pre- and post-composition with a multiplication by a nonzero constant.

We approach the discussion of covering properties of $\mathbf{h}_{f_0}$ indirectly, by introducing a different, more convenient, dynamical model.

By the Riemann Mapping Theorem, there is a conformal isomorphism ψ from B_0 to the cut plane $\mathbb{C}\backslash[0, \infty)$ sending $-1/2$ (the critical point of f_0) to -1. These conditions do not fix ψ uniquely, but it is not difficult to see that ψ can be chosen so that $\psi'(-1/2) > 0$ and, under this condition, becomes unique. Because of uniqueness, and because B_0 and the cut plane are invariant under complex conjugation, ψ commutes with complex conjugation. It must therefore map $(-1, 0)$ to $(-\infty, 0)$, these being respectively the real points of B_0 and those of the cut plane. Since $\psi'(-1/2) > 0$, ψ is strictly increasing on $(-1, 0)$, and $\lim_{x \to 0^-} \psi(x) = 0$.

Define an analytic mapping K of the cut plane to itself by

$$K = \psi \circ f_0 \circ \psi^{-1}. \tag{3.5}$$

So defined, K is a real-symmetric analytic degree-2 branched covering:

$$K : \mathbb{C}\backslash[0, \infty) \to \mathbb{C}\backslash[0, \infty).$$

By construction, -1 is a critical point of h, and the corresponding critical value is $\psi(-1/4)$. There is a simple explicit formula for h.

Proposition 3.15 *The mapping $K(z)$ is the Koebe function*

$$K(z) = \frac{z}{(z-1)^2}. \tag{3.6}$$

Proof For $x \in [0, \infty)$ and small positive ε, $K(x + i\varepsilon)$ and $K(x - i\varepsilon)$ are complex conjugates. As $\varepsilon \to 0^+$, $x + i\varepsilon$ converges to the boundary of the cut plane, so $\psi^{-1}(x + i\varepsilon)$ converges to the boundary of B_0, so $f_0(\psi^{-1}(x + i\varepsilon))$ also converges to the boundary of B_0, so $K(x + i\varepsilon)$ converges to the boundary $[0, \infty]$ of the cut plane.

We apply the Schwarz reflection to see that K can be extended analytically through $[0, \infty)$. The above argument does not rule out that $K(x + i\varepsilon)$ goes to ∞. To get around this, we recall that K is a degree-2 branched covering, and that, on a small disk about the critical point -1, it takes on every value near the critical value $K(-1)$ twice. Hence, $K(x + i\varepsilon)$ stays away from $K(-1)$, so Schwarz reflection can be applied without difficulty to the function $1/(K(-1) - K(z))$, which is bounded on a neighborhood of the positive axis.

Thus, K extends to a meromorphic function on the finite plane. Since $1/(K(-1) - K(z))$ is bounded near ∞, K is also meromorphic at ∞, and hence, it is a rational function. Since it takes each finite nonreal value exactly twice, it has degree 2. It never takes on a value in $[0, \infty]$ anywhere off $[0, \infty]$ so it must map $[0, \infty]$ to itself, with each point having two preimages, counted with multiplicity. As x runs up the positive axis, $K(x)$ starts at 0 and runs up, reaching ∞ at some finite a; it then runs back down, approaching 0 as $x \to \infty$. The pole at a must have order 2, so $K(z)$ must have the form

$$K(z) = (bz^2 + cz + d)/(z - a)^2.$$

From $K(0) = K(\infty) = 0$, we have $b = d = 0$. Thus,

$$K(z) = cz/(z - a)^2$$

for some constant c, which clearly must be positive. By a simple calculation, a function of this form has only one (finite) critical point at $-a$, so $a = 1$. It remains only to show $c = 1$.

We saw above, from elementary considerations that $\psi(x) \to 0$ as $x \to 0^-$. Hence, for small negative x,

$$K^n(\psi(x)) = \psi(f_0^n(x)) \to 0,$$

so there are at least some K-orbits converging to 0; this rules out $c > 1$, so we need only show that $c < 1$ is also impossible.

By Theorem 3.2, and the Carathéodory Theorem, ψ has a continuous extension to the boundary $J(f_0)$ of B_0, which we continue to denote by ψ. The extension is not a homeomorphism; it identifies complex conjugate pairs on $J(f_0)$. A useful invertibility property can be formulated as follows: $J(f_0) \backslash -1, 0$ is a disjoint union of two Jordan arcs, and the extended ψ maps each of these arcs homeomorphically to $(0, \infty)$. Note that the conjugation equation, written as

$$K \circ \psi = \psi \circ f,$$

extends by continuity to $\overline{B_0}$. Because f_0 on $J(f_0)$ is conjugate to $s \mapsto 2s$ on $\mathbb{T}$, there exist infinite backward orbits for f_0 in $J(f_0)$ which converge to 0, i.e., sequences $z_0, z_{-1}, z_{-2}, \ldots$ with $z_{-n+1} = f(z_{-n})$ and $z_{-n} \to 0$. Then $\psi(z_{-n}) \to 0$ and $K(\psi(z_{-n})) = \psi(z_{-n+1})$, that is, $(\psi(z_{-n})$ is a backward orbit for K converging to 0. This rules out $c < 1$, so the only remaining possibility is $c = 1$, so the formula (3.6) is proved.   □

We note several other contexts in which the Koebe function (3.6) comes up. It is, of course, the essentially unique univalent function on the unit disk which saturates the Koebe distortion estimates. (See, for example, [Co], Theorem 7.9 and p. 31.)

Furthermore, and perhaps more instructively for our study, one can deduce from the above discussion that $K(z)$ is the *conformal mating* (see [Do1]) of the map f_0 with the Chebysheff quadratic polynomial $f_{-2}(z) = z^2 - 2$.

We next develop an important piece of technique:

Proposition 3.16 *Let f_1, f_2 each have a normalized simple parabolic point at 0, and assume that their restrictions to their respective principal basins are conformally* *conjugate, i.e., that there is a conformal isomorphism* $\varphi : B_0^{f_1} \to B_0^{f_2}$ *so that*

$$f_1 = \varphi^{-1} \circ f_2 \circ \varphi \quad \text{on } B_0^{f_1}. \tag{3.7}$$

Assume further that φ *and* φ^{-1} *extend continuously to the boundary point* 0 *of* $B_0^{f_1}$
respectively $B_0^{f_2}$. *Then there is a conformal isomorphism* $\psi : W_{f_1}^+ \to W_{f_2}^+$, *sending*
0 *to* 0, *and a nonzero constant* v *so that*

$$h_{f_1}^+ = v \cdot h_{f_2}^+ \circ \psi.$$

We begin with a few simple remarks. From the continuity of φ at 0, it follows that
$\varphi(0) = 0$,

$$\varphi(0) = \lim_{n \to \infty} \varphi(f_1^n(z)) = f_2^n(\varphi(z)) = 0 \quad \text{for any } z \in B_0^{f_1}.$$

It is easy to verify, again making use of continuity of φ at 0, that φ maps attracting
petals for f_1 to attracting petals for f_2, and that φ^{-1} maps attracting petals for f_2 to
attracting petals for f_1. For the proposition to make sense, we have to have chosen
normalizations for attracting and repelling Fatou coordinates for the two mappings,
but the choices are obviously immaterial. We fix, however we like, the normalization
of $\phi_A^{(2)}$. Then $\phi_A^{(2)} \circ \varphi$ is an attracting Fatou coordinate for f_1, so by the uniqueness
of Fatou coordinates (Proposition 2.11), $\phi_A^{(2)} \circ \varphi - \phi_A^{(1)}$ is constant. For notational
simplicity, we may assume

$$\phi_A^{(1)} = \phi_A^{(2)} \circ \varphi;$$

this has the effect of making the constant v in the proposition equal to one.

The situation with repelling Fatou coordinates is more complicated. It is not true
that φ maps repelling petals to repelling petals, since φ is only available on $B_0^{f_1}$ and
repelling petals are not contained in B_0. What we have to work with instead is the
weaker observation that φ maps f_1-orbits coming out from 0 in $B_0^{f_1}$ to the same kind
of orbits for f_2. To exploit this observation, we need to develop some technique for
a single function with a simple parabolic fixed point.

Let f have the form (2.1); let P_R be a repelling petal for f, and write f^{-1} for the
inverse of the restriction of f to P_R. We define

$$\widetilde{P_R} := \{z \in P_R : f^{-n}(z) \in B_0^f \text{ for } n = 0, 1, \ldots\}.$$

In the usual way, it is straightforward to show that the image under $\mathrm{ixp} \circ \phi_R$ of $\widetilde{P_R}$
is independent of the petal P_R. We denote this image by $\widetilde{W}$; since $\widetilde{P_R} \subset P_R \cap B^f$,
$\widetilde{W}$ is contained in $\mathscr{D}(\mathbf{h}_f)$.

Lemma 3.4

(1) Let $t \mapsto \tau(t), 0 \le t \le 1$ be a continuous arc in $P_R \cap B^f$ with $\tau(0) \in \tilde{P}_R$. Then $\tau(1) \in \tilde{P}_R$.
(2) $\tilde{P}_R$ is open (so its image $\tilde{W}$ under ixp $\circ \phi_R$ is also open).
(3) A connected component of $\tilde{W}$ is also a connected component of $\mathcal{D}(\mathbf{h})$. In other words, a component of $\mathcal{D}(\mathbf{h})$ is either disjoint from $\tilde{W}$ or contained in it.
(4) If there is a continuous path τ in $P_R \cap B_0^f$ from z to $f^{-1}(z)$, then $z \in \tilde{P}_R$.
(5) The components $W^+\backslash\{0\}$ and $W^-\backslash\{\infty\}$ of $\mathcal{D}(\mathbf{h})$ are contained in $\tilde{W}$.

Proof

(1) If $\tau(1) \notin \tilde{P}_R$, there must be an n so that $f^{-n}(\tau(1)) \notin B_0^f$. Fix such an n, and let

$$t_0 := \inf\{t : f^{-n}(\tau(t)) \notin B_0^f\}.$$

Since $\tau(.)$ and f^{-1} are continuous and B_0^f is open,

$$f^{-n}(\tau(t_0)) \in \partial B_0^f \subset \partial B^f,$$

so in particular $f^{-n}(\tau(t_0)) \notin B^f$. But B^f is backward-invariant under f and the path τ is by assumption in B^f, so this is a contradiction.
(2) If $z_0 \in \tilde{P}_R$ and z is near enough to z_0, then the straight-line segment $[z_0, z]$ is in $P_R \cap B^f$, and so, by 1), in $\tilde{P}_R$.
(3) Since $\tilde{W}$ is contained in $\mathcal{D}(\mathbf{h})$, each component of $\tilde{W}$ is contained in a component of $\mathcal{D}(\mathbf{h})$. Let U be a component of $\tilde{W}$ and V the component of $\mathcal{D}(\mathbf{h})$ which contains it. If U is not all of V, the relative boundary must be nonempty, i.e., there must exist a $w_0 \in (\partial U) \cap V$. This w_0 is in $\mathcal{D}(\mathbf{h})$, so it can be written as $w_0 = \mathrm{ixp} \circ \phi_R(z_0)$ with $z_0 \in P_R \cap B^f$. Let D be an open disk about z_0, small enough to be contained in $P_R \cap B^f$ and also so that ixp $\circ \phi_R$ maps it homeomorphically into V. Since ixp $\circ \phi_R(D)$ contains points of $\tilde{W}$, D contains points of $\tilde{P}_R$, and it then follows from 1) that $D \subset \tilde{P}_R$, so $z_0 \in \tilde{P}_R$, so $w_0 \in \tilde{W}$, contradiction.
(4) We argue by induction on n that $f^{-n}\tau \subset B_0^f$ for all n. This is true by hypothesis for $n = 0$. Suppose it is true for n but not for $n + 1$. Then $f^{-(n+1)}(\tau(0)) = f^{-n}(\tau(1)) \in B_0^f$, but $f^{-(n+1)}(\tau(t_0)) \in \partial B_0^f$ for some t_0. Then $f^{-(n+1)}(\tau(t_0)) \notin B^f$, which contradicts $\tau(t_0) \in B^f$ by backward invariance of B^f. The contradiction proves $f^{-n}\tau \subset B_0^f$ for all n, so in particular, $f^{-n}(\tau(0)) = f^{-n}(z) \in B_0^f$ for all n, i.e., $z \in \tilde{P}_R$.
(5) Since the resulting domain $\tilde{W}$ does not depend on the repelling petal used to construct it, we may assume that P_R is ample. Then a sector

$$\{z : 0 < |z| < \rho, \pi/2 - \delta < \mathrm{Arg}(z) < \pi/2 + \delta\},$$

with ρ and δ small enough, is contained in $P_R \cap B_0^f$. By (4) there is a sector of this form but with smaller δ and ρ which is contained in $\widetilde{P}_R$, and by the proof of Proposition 2.3, the image under ixp $\circ \phi_R$ of this smaller sector is a punctured neighborhood of 0. Thus, $\widetilde{W}$ intersects $W^+\backslash\{0\}$, so by 3), $W^+\backslash\{0\}$ is one of the components of $\widetilde{W}$. The proof for $W^-\backslash\{\infty\}$ is similar. $\square$

We return to the proof of Proposition 3.16; we are now ready to construct the mapping ψ. We fix repelling petals $P_R^{(1)}$ for f_1 and $P_R^{(2)}$ for f_2. The situation is the familiar one: We need these petals to make a construction, but the objects constructed turn out not to depend on them. We let $\phi_R^{(1)}$ be a repelling Fatou coordinate for f_1 defined at least on $P_R^{(1)}$, and similarly $\phi_R^{(2)}$ for f_2. Let $w_0 \in \widetilde{W}_1$, and write $w_0 = \text{ixp} \circ \phi_R^{(1)}(z_0)$ with $z_0 \in P_R^{(1)}$. Then $z_{-n} := f^{-n}(z_0)$ is an outcoming orbit in B_0^f. Because φ conjugates f_1 to f_2 (on their respective principal basins) and is continuous at 0, $y_{-n} = \varphi(z_{-n})$ is an outcoming orbit for f_2. For sufficiently large n, y_{-n} is in $P_R^{(2)}$. Take any n_0 so that $y_{-n} \in P_R^{(2)}$ for $n \geq n_0$, and set

$$\psi(w_0) = \text{ixp} \circ \phi_R^{(2)}(y_{-n_0}) \in \widetilde{W}_2.$$

It is easy to check that the result does not depend on the choices of repelling petals, preimage z_0, or number n_0 of steps back before applying ixp $\circ \phi_R^{(2)}$, so we have constructed a mapping from $\widetilde{W}_1$ to $\widetilde{W}_2$. Each of the steps

$$w_0 \mapsto z_0 \mapsto z_{-n_0} = f^{-n_0}(z_0) \mapsto y_{-n_0} \mapsto \psi(w_0) = \text{ixp} \circ \phi_R^{(2)}(y_{-n_0})$$

can be done in a neighborhood of each of the points involved by applying a local conformal isomorphism, so $\psi(.)$ is a local conformal isomorphism. Carrying out the same construction with f_1 and f_2 interchanged and φ^{-1} in the role of φ produces a mapping from $\widetilde{W}_2$ to $\widetilde{W}_1$ which inverts ψ, so ψ is a global conformal isomorphism. Furthermore,

$$\mathbf{h}_2(\psi(w_0)) = \text{ixp} \circ \phi_A^{(2)}(y_{-n_0}) = \text{ixp} \circ \phi_A^{(2)}(\varphi(y_{-n_0}))$$
$$= \text{ixp} \circ \phi_A^{(1)}(z_{-n_0}) = \text{ixp} \circ \phi_A^{(1)}(z_0) = \mathbf{h}_1(w_0).$$

The mapping ψ just constructed is a more global version of the one in the proposition. To complete the proof of the proposition, we need to show that

1. ψ maps $W_1^+\backslash\{0\}$ to $W_2^+\backslash\{0\}$.
2. ψ extends analytically to map 0 to 0

Proof of (1) Since $\psi : \widetilde{W}_1 \to \widetilde{W}_2$ is a homeomorphism, it maps components to components. We know from 3 of Lemma 3.4 that $W_1^+\backslash\{0\}$ is a component of $\widetilde{W}_1$; we have only to show that the component it maps to is $W_2^+\backslash\{0\}$. To do this, choose ample petals $P_A^{(1)}$ and $P_R^{(1)}$ for f_1, and let $t \mapsto \lambda_1(t)$ be an arc in $P_A^{(1)} \cap P_R^{(1)}$, ending at 0, which is mapped by $\phi_A^{(1)}$ to an upward vertical ray. By the asymptotic estimate for Fatou coordinates, λ projects under $ixp \circ \phi_R^{(1)}$ to an arc $\tilde{\lambda}_1$ in $\widetilde{W}_1$ ending at 0. Let $\lambda_2 := \varphi \circ \lambda_1$. It is then immediate that λ_2 maps under $\phi_A^{(2)}$ to an upward vertical ray.

Claim λ_2 *is tangent to the positive imaginary axis at* 0.

Accepting the claim for the moment: We can then project λ_2 under $ixp \circ \phi_R^{(2)}$ to an arc $\tilde{\lambda}_2$ in $\widetilde{W}_2$. By the asymptotic estimates on Fatou coordinates, $\tilde{\lambda}_2$ ends at 0. By the construction of ψ,

$$\tilde{\lambda}_2 = \psi \circ \tilde{\lambda}_1.$$

Thus: $\psi(W_1^+ \backslash \{0\})$ intersects W_2^+ and hence by connectedness,

$$\psi(W_1^+ \backslash 0) = W_2^+ \backslash \{0\}.$$

Proof of claim It would be immediate if we knew that a terminal subarc of λ_2 is contained in an attracting petal. Rather than prove this directly, we argue as follows. Let $P_A^{(2)}$ be an attracting petal for f_2 which maps under $\phi_A^{(2)}$ to a right half-plane. Since λ_2 is contained in B^f, there is an n so that $f_1^n(\lambda_s(0)) \in P_A^{(2)}$. Since $\text{Re}(\phi_A^{(2)})$ is constant on λ_2, $\lambda_2(t)$ stays in $P_A^{(2)}$, and as already noted approaches 0 as $t \to 1^-$. By the crude estimates on Fatou coordinates, $f_2^n \circ \lambda_2$ is tangent to the positive imaginary axis at 0. Applying an analytic local inverse of f_2 n times proves the claim.

Proof of (2) This can be proved in a straightforward way using crude local estimates on Fatou coordinates, but it is quicker to argue that

$$\psi = (h_2^+)^{-1} \circ h_1^+ \quad \text{on a punctured neighborhood of } 0,$$

(where $(h_2^+)^{-1}$ means an analytic local inverse of h_2^+, which exists since h_2^+ has a simple zero at 0) and to note that the right-hand side is analytic at 0. To prove the preceding equation, we invoke the general formula $h_1^+ = h_2^+ \circ \psi$ and note that by (1), ψ and $(h_2^+)^{-1} \circ h_1^+$ agree at points of the form $\tilde{\lambda}_1(t)$ for t near enough to 1. The assertion then follows by analytic continuation. The proof of Proposition 3.16 is thus completed. $\square$

Definition 3.2 For K as in (3.6), let us select a real symmetric Fatou coordinate ϕ_R^K, and a Fatou coordinate ϕ_A^K, so that

$$h_K'(0) = 1 \quad \text{and} \quad \mathscr{D}(h_K^+) = \mathbb{D}, \quad \mathscr{D}(h_K^-) = \mathbb{C}\backslash\mathbb{D}, \quad \text{where } \mathbf{h}_K = (h_K^+, h_K^-).$$

We will denote this normalized map $\mathbf{h}_K$ by $\mathbf{k} = (k^+, k^-)$ for the remainder of this book.

In view of the above discussion, we note:

Proposition 3.17 *There exist $v \neq 0$ and $\theta \in \mathbb{R}$ such that*

$$\mathbf{h}_{f_0} = v \cdot \mathbf{k} \circ \psi(e^{2\pi i\theta}z), \quad \text{where } \psi : \mathbb{D} \rightarrow \hat{W}^+$$

is the conformal Riemann mapping, which satisfies $\psi(0) = 0$, $\psi'(0) > 0$.

We show the following.

Theorem 3.3 $\mathbf{h}_{f_0}$ *has*

- *exactly one critical value $v := \mathrm{ixp} \circ \phi_A(-1/4)$;*
- *exactly two asymptotic values 0 and ∞.*

Proof To show that the above-defined v is the only possible critical value of $\mathbf{h}_{f_0}$, we write

$$\mathbf{h}_{f_0} = \left(\mathrm{ixp} \circ \phi_A\right) \circ \left(\mathrm{ixp} \circ \phi_R\right)^{-1},$$

for an appropriate branch of the inverse. Whatever inverse branch is used, its derivative does not vanish. Hence, a critical value of $\mathbf{h}_{f_0}$ is a critical value $(\mathrm{ixp} \circ \phi_A)$, i.e., the image under $\mathrm{ixp}(.)$ of a critical value of ϕ_A. Since $-1/4$ is the unique critical value of f_0, it follows from Proposition 2.12 that the critical values of ϕ_A have the form $\phi_A(-1/4) - n$, $n \geq 1$. Since $\mathrm{ixp}(.)$ is 1-periodic, v is the unique critical value of $\mathrm{ixp} \circ \phi_A$.

We turn to determining the asymptotic values of $\mathbf{h}_{f_0}$. We show that the asymptotic values of the restriction $\mathbf{h}_{f_0}|_{\hat{W}^+}$ are $\{0, \infty\}$; the argument for $\mathbf{h}_{f_0}|_{\hat{W}^-}$ is identical. To simplify geometric considerations, we invoke Proposition 3.17, and present the proof for $\mathbf{k}$.

We need a number of elementary facts about K which we summarize below for ease of reference:

Proposition 3.18 *The Koebe function $K(z) = z/(1 - z)^2$ is a degree-2 ramified cover $\hat{\mathbb{C}} \rightarrow \hat{\mathbb{C}}$; its critical values are $-1/4$ and ∞; the corresponding critical points are -1 and $+1$. The interval $[0, \infty]$ is fully invariant under h; the interval $(-\infty, 0]$ is forward invariant.*

The interval $[-\infty, -1/4]$ contains both of the critical values of h, so h is a cover in the strict sense above $\mathbb{C}\backslash(-\infty, -1/4]$. In other words: h admits two analytic right-inverses defined on the cut plane $\mathbb{C}\backslash(-\infty, -1/4]$. One of these inverse branches sends 0 to 0; the other sends 0 to its other preimage ∞. At the cost of introducing a little ambiguity, we will write simply K^{-1} for the inverse branch which is regular at 0. A straightforward elementary calculation shows that the preimage of $(-\infty, 0]$ under h is the unit circle, so the image of K^{-1} is the unit disk $\mathbb{D}$.

Proposition 3.19 *The inverse branch K^{-1} maps the open upper (respectively lower) half-plane into itself. It also maps $\mathbb{C}\backslash(-\infty, 0]$ into itself. The sequence of iterates $(K^{-1})^n$ converges uniformly to 0 on compact subsets of $\mathbb{C}\backslash(-\infty, 0]$. Any repelling Fatou coordinate extends to a univalent analytic function on $\mathbb{C}\backslash(-\infty, 0]$.*

Proof The proof may be given as an application of the Denjoy-Wolff Theorem, however, a more direct argument is possible in our case.

Elementary considerations show that the upper half-plane is mapped into itself by K^{-1}, and similarly for the lower half-plane. Furthermore,

$$K^{-1} : \mathbb{C}\backslash(-\infty, 0] \longrightarrow \mathbb{C}\backslash(-\infty, 0].$$

By Montel's theorem, the sequence of iterates $(K^{-1})^n$ is a normal family on $\mathbb{C}\backslash(-\infty, 0]$. Elementary considerations show that $(K^{-1})^n(x) \to 0$ for x real, positive, and small. By Vitali's theorem, $(K^{-1})^n$ converges uniformly to 0 on compact subsets of $\mathbb{C}\backslash(-\infty, 0]$.

The preceding assertion says that

$$B_0^{K^{-1}} = \mathbb{C}\backslash(-\infty, 0].$$

Note that any K^{-1}-orbit starting in $(-1/4, 0)$ gets in finitely many steps into $(-1, -1/4]$, which is outside $\mathscr{D}(K^{-1})$. Thus, $(-1/4, 0)$, although in $\mathscr{D}(K^{-1})$, is outside of $B^{K^{-1}}$. Using repeatedly the functional equation

$$\phi_R(K^{-1}(z)) = \phi_R(z) - 1,$$

we see that any repelling Fatou coordinate extends (uniquely) to a function defined on all of $\mathbb{C}\backslash(-\infty, 0]$ and satisfying the above functional equation everywhere. The extended function is analytic; from the univalence of K^{-1} and the fact that repelling Fatou coordinates are univalent on small petals, the extended function is also univalent. □

Next we introduce, in the context of the particular mapping K, a useful dynamically-defined family of arcs which come up, with minor variations, at a number of places in this work. Let α be a real number so that

$$\text{Re}(\phi_A(-1)) < \alpha < \text{Re}(\phi_A(-1/4)) \, (= \text{Re}(\phi_A(-1)) + 1). \tag{3.8}$$

Proposition 3.20 *There is an attracting petal P which maps under ϕ_A to the right half-plane $\{u + iv : u > \alpha\}$.*

For α a sufficiently large real number, this proposition is part of the local theory of parabolic fixed points. Showing that in the present situation, any $\alpha > \text{Re}(\phi_A(-1))$ is large enough depends on some covering properties of K. We omit the proof here.

The boundary of P can be written as $\{0\} \cup \gamma$, where γ is an arc in the cut plane $\mathbb{C}\backslash[0, \infty)$ which approaches the parabolic point 0 at both ends. The arc γ is a component of the preimage under ϕ_A of the vertical line $\{\alpha + iv : -\infty < v < \infty\}$; we give it the counterclockwise orientation, corresponding to the decreasing orientation for the vertical line: $\text{Im}(\phi_A(z))$ runs from $+\infty$ to $-\infty$ as z traverses γ in the counterclockwise sense. This can be seen using the crude asymptotic formula $\phi_A(z) \approx -1/z$, valid for small z not too near to the positive real axis.

We will speak of an arc $s \mapsto \sigma(s)$, defined and continuous for $0 < s < 1$, i.e., without beginning or end point, as an *open* arc. We say that such an arc *starts at* z_0 if $\lim_{s \to 0^+} \sigma(s) = z_0$ and that it *ends at* z_1 if $\lim_{s \to 1^-} \sigma(s) = z_1$.

Proposition 3.21 *With γ as above:*

1. *the preimage of γ under K^n is a disjoint union of 2^n smooth arcs in the cut plane. Each of the component arcs starts from and ends at a preimage of the parabolic point 0. The complex conjugate of a component arc of $K^{-n}\gamma$ is a component arc;*
2. *the component arcs of $K^{-n}\gamma$ do not intersect the real axis; each of them lies either entirely in the upper half-plane or entirely in the lower half-plane;*
3. *one of the component arcs of the preimage of γ under K, equipped with the pullback of the orientation of γ under h, lies in the upper half-plane, starts at 0, and ends at ∞. The other component is the complex conjugate of this one, with its orientation reversed. It lies in the lower half-plane and runs from ∞ to 0.*

Proof For a postcritical point $z = K^n(-1)$, $n \geq 0$,

$$\mathrm{Re}(\phi_A(z)) = \mathrm{Re}(\phi_A(K^n(-1))) = \mathrm{Re}(\phi_A(-1)) + n \neq \alpha,$$

by the condition (3.8) on α. Hence, the postcritical set does not intersect γ, so K^n is a strict cover of degree 2^n over γ, i.e., $(K^n)^{-1}\gamma$ has 2^n components, each of which is a smooth open arc. Since γ starts from and ends at 0, each component starts from and ends at a preimage, of order $\leq n$, of 0 (Fig. 3.2).

Since K commutes with complex conjugation, $\mathrm{Re}(\phi_A(\bar{z})) = \mathrm{Re}(\phi_A(z))$. Hence, the image of P under complex conjugation, which is again a petal, is mapped by ϕ_A

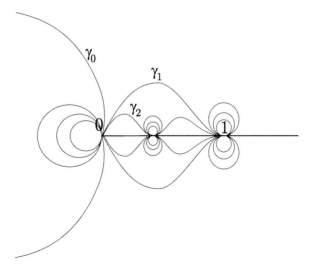

Fig. 3.2 A few preimages of γ under the iteration of K (Proposition 3.21)

to the same right half-plane as P, so P is invariant under complex conjugation, so its boundary γ is also invariant. Again, using the fact that K commutes with complex conjugation, the preimage under K^n of γ is invariant, so the complex conjugate of each component arc is a component arc.

To show that preimages of γ do not intersect $\mathbb{R}$, it is easy to show, using the estimate $\phi_A(z) \approx -1/z$, that $x \mapsto \mathrm{Re}(\phi_A(x))$ is strictly increasing for x small and negative. Using $\phi_A(x) = \phi_A(K^n(x)) - n$, and the fact that all iterates K^n are strictly increasing on $(-1, 0)$, we see that $\mathrm{Re}(\phi_A(x))$ is strictly increasing on all of $(-1, 0)$. On the other hand, K is strictly decreasing on $(-\infty, -1)$ and maps this interval onto $(-1/4, 0)$, so, again applying $\phi_A(x) = \phi_A(K(x)) - 1$, $\mathrm{Re}(\phi_A(x))$ is strictly decreasing on $(-\infty, -1)$. In particular,

$$\mathrm{Re}(\phi_A(x)) \geq \mathrm{Re}(\phi_A(-1)) \quad \text{for } x \in (-\infty, 0).$$

If $K^n(z) \in \gamma$ (with $n > 0$),

$$\mathrm{Re}(\phi_A(z)) = \mathrm{Re}(\phi_A(K^n(z))) - n = \alpha - n$$
$$\leq \alpha - 1 < \mathrm{Re}(\phi_A(-1)) = \min_{x<0} \mathrm{Re}(\phi_A(x)).$$

Thus, the preimage under K^n of γ does not intersect the negative real axis. Since γ does not intersect $[0, \infty]$, which is fully invariant, the preimage does not intersect it either, and hence does not intersect $\mathbb{R}$.

By condition (3.8) on α, the critical value $-1/4$ of K is in P, so the Jordan curve made by appending 0 to γ, the boundary of P, runs once around the critical value. Hence, lifts of γ under K do not close. One of the two lifts starts at 0; it cannot end at 0 and so must end at ∞, which is the other preimage of 0. Since this lift starts in the upper half-plane and, by assertion (2), does not intersect the real axis, it must lie in the upper half-plane.   □

We denote by

- γ_0 the component of $K^{-1}\gamma$ starting at 0, i.e., the component in the upper half-plane;
- γ_1 the component of $K^{-1}\gamma_0$ which begins at 0. Then γ_1 also lies in the upper half-plane and ends at 1, the preimage of ∞;
- γ_2 the component of $K^{-1}\gamma_1$ which starts at 0. Again, γ_2 lies in the upper half-plane; it ends at the unique preimage x_2 of 1 in $(0, 1)$, whose value can easily be computed to be $(3 - \sqrt{5})/2$.

Then $\{0\} \cup \gamma_1 \cup \{1\} \cup \overline{\gamma_1}$ is a Jordan curve; we denote its interior by P_R.

Lemma 3.5 P_R is an attracting petal for K^{-1}.

Proof Since

- $\partial(K^{-1}P_R) = K^{-1}(\partial P_R) = \{0\} \cup \gamma_2 \cup \{x_2\} \cup \overline{\gamma_2}$,
- $x_2 \in (0, 1) \subset P_R$,
- $\gamma_2 \cup \overline{\gamma_2}$ does not intersect ∂P,

it follows that the closure of $K^{-1}P_R$ is contained in $P_R \cup \{0\}$. By local theory, γ_1 is tangent to the imaginary axis at 0; from this it follows that every K^{-1}-orbit with initial point not in $(-\infty, 0]$ is eventually in P_R.

It remains only to verify condition (4) of Definition 2.1. We showed above, using Vitali's theorem, that $(K^{-1})^n$ converges uniformly to 0 on compact subsets of $\mathbb{C}\backslash(-\infty, 0]$. Since ∂P_R is tangent to the imaginary axis at 0, it follows from local theory that $(K^{-1})^n$ converges uniformly to 0 on the intersection of P_R with a sufficiently small disk centered at 0; uniform convergence on all of P_R follows. □

Note the mixed character of P_R: although it is a repelling petal, its boundary is defined as a curve with constant $\mathrm{Re}(\phi_A)$, except at the two points 0 and x_2, where the ϕ_A is not defined.

By Proposition 2.9, $C_R := P_R \backslash K(P_R)$ projects bijectively to the repelling cylinder $\mathscr{C}_R$ and hence is mapped bijectively by ixp $\circ \phi_R$ to the punctured plane $\mathbb{C}\backslash\{0\}$. We write C_R^+ and C_R^- for the intersections of C_R with the open upper and lower half-planes respectively; then

$$C_R = C_R^+ \cup (x_2, 1) \cup C_R^-.$$

ixp $\circ \phi_A$ is analytic on C_R^+ and on C_R^- but cannot be continued analytically through any point of $(x_2, 1)$. Thus, the image under ixp $\circ \phi_R$ of C_R^+ (respectively C_R^-) is W^+ (respectively W^-). By an argument given above for ϕ_A,

$$\phi_R(\bar{z}) = \overline{\phi_R(z)} + c \quad \text{with } c \text{ a pure imaginary constant,}$$

so the imaginary part of ϕ_R is constant on $(0, \infty)$, and hence ixp $\circ \phi_R$ maps $[r_2, 1]$ to a circle centered at the origin. We chose the real part of the additive constant in ϕ_R so that ϕ_R is real on $(0, \infty)$; then

$$W^+ = \mathbb{D}\backslash\{0\} \quad \text{and} \quad W^- = \mathbb{C}\backslash\overline{\mathbb{D}}.$$

We now have all the pieces in place to prove that the only asymptotic values of $\mathbf{k}$ are 0 and ∞. We assume that $s \mapsto \tau(s), s \in [0, 1)$ is a continuous arc in W^+ converging to the unit circle as $s \to 1$ (but not necessarily to an particular point of the circle) such that $\mathbf{k}(\tau(s))$ converges as $s \to 1$ to a finite nonzero value z_∞, and we show this leads to a contradiction. We select an α as in (3.8) and also so that

$$\frac{1}{2\pi}\mathrm{Arg}(z_\infty) - \alpha \notin \mathbb{Z}.$$

Since $\mathbf{k}(\tau(s)) \to z_\infty$, we can, by deleting an initial segment of τ and reparametrizing, assume that

$$\frac{1}{2\pi}\mathrm{Arg}(H(\tau(s))) - \alpha \notin \mathbb{Z} \quad \text{for } 0 \le s < 1. \tag{3.9}$$

Recall that ixp $\circ$ ϕ_R maps C_R^+ bijectively to W^+. We denote by $\hat{\tau}(s)$ the image of $\tau(s)$ under the inverse of this bijection. We claim that the lift $\hat{\tau}(.)$ is continuous. To see this, we note that

$$\mathbf{k}(\tau(s)) = \exp(2\pi i\phi_A(\hat{\tau}(s))),$$

so by (3.9).

$$\mathrm{Re}(\phi_A(\hat{\tau}(s))) - \alpha \notin \mathbb{Z} \quad \text{for } 0 \leq s < 1, \tag{3.10}$$

that is, the values of $\hat{\tau}(s)$ lie in the interior of P_R^+, not on the "edges". Since ixp$\circ\phi_R$ is a local homeomorphism everywhere in the interior of C_R^+, continuity of $\hat{\tau}(s)$ follows.

We recall that P denotes here the attracting petal mapped by ϕ_A to the half-plane $\{\mathrm{Re}(w) > \alpha\}$. Since the orbit of the initial point $\hat{\tau}(0)$ of the lift is eventually in P, and since $\mathrm{Re}(\phi_A(\hat{\tau}(0))) - \alpha \notin \mathbb{Z}$, there exist positive integers n and m such that

$$K^n(\hat{\tau}(0)) \in C := \{z \in P : \alpha + m < \mathrm{Re}(\phi_A(z)) < \alpha + m + 1\}.$$

Because of (3.10), $\hat{\tau}(s)$ never intersects either boundary arc of the crescent C, and so by continuity,

$$\tilde{\tau}(s) := K^n(\hat{\tau}(s)) \in C \quad \text{for } 0 \leq s < 1.$$

Because the base arc $\tau(s)$ (in W^+) approaches the unit circle as $s \to 1$, $\hat{\tau}(s)$ approaches $[x_2, 1]$, so because $[0, \infty]$ is mapped to itself by K, $\tilde{\tau}(s)$ approaches $[0, \infty]$ as $s \to 1$. We thus have the following situation:

1. $s \mapsto \tilde{\tau}(s)$ is a continuous arc in the crescent C.
2. $s \mapsto \tilde{\tau}(s)$ approaches $[0, \infty]$ as $s \to 1$.
3. $\mathrm{ixp}(\phi_A(\tilde{\tau}(s)))$ approaches a finite nonzero limit z_0 as $s \to 1$.

It is now easy to show that these three assertions lead to a contradiction. We first argue that (1) and (2) imply

$$|\mathrm{Im}(\phi_A(\tilde{\tau}(s)))| \to \infty. \tag{3.11}$$

Then $\mathbf{k}(\tau(s)) = \exp(2\pi i\phi_A(\tilde{\tau})(s))$ converges to 0 (if $\mathrm{Im}(\phi_A(\tilde{\tau})(s)) \to +\infty$) or to ∞, contradicting (3).

To prove (3.11), for sufficiently small positive ε, there is an attracting petal P_ε mapped homeomorphically by ϕ_A to the half-plane $\{\mathrm{Im}(w) > \alpha - \varepsilon\}$. Let r be a positive real number and let R_r denote the preimage under the restriction of ϕ_A to such a P_ε of the compact rectangle

$$\{u + iv : \alpha + m \leq u \leq \alpha + m + 1, -r \leq v \leq r\}.$$

Since R_r is compact and disjoint from $[0, \infty]$, its distance from $[0, \infty]$ is strictly positive. By (1), if $|\mathrm{Im}(\phi_A(\tilde{\tau}(s)))| \leq r$, then $\tilde{\tau}(s) \in R_r$. Hence, by (2), $|\mathrm{Im}(\phi_A(\tilde{\tau}(s)))|$ is eventually $> r$. Since this is true for all r, (3.11) holds.

It is evident that $\{0, \infty\} \subset \mathrm{Asym}(H)$. Indeed, for small enough $\varepsilon > 0$, the projection of $\gamma_1 \cap D_\varepsilon(1)$ by $\mathrm{ixp} \circ \phi_R$ is a simple curve in $\mathbb{D}$ whose image under H lands at 0; similarly γ_2 yields a curve whose image under H lands at ∞.

Now let

$$\tau : [0, 1) \to \mathbb{D} = \mathscr{D}(H)$$

be a curve such that $\tau(t)$ converges to the unit circle $\mathbb{T}$ as $t \to 1-$. Let

$$\tilde{\tau} : [0, 1) \subset \hat{\mathbb{C}} \backslash \mathbb{R}_{\geq 0}$$

be a component of its lift under $\mathrm{ixp} \circ \phi_R$, parametrized so that

$$\tilde{\tau}(t) \to \mathbb{R}_{\geq 0} \text{ as } t \to 1-. \tag{3.12}$$

Let C' be a connected component of $K^n(C)$ for $n \in \mathbb{Z}$. We will say that the curve $\tilde{\tau}$ crosses C' if there exist $t_1 \neq t_2 \in (0, 1)$ and $n \in \mathbb{Z}$ such that

$$\tilde{\tau} : (t_1, t_2) \to C', \quad K^n(\tilde{\tau}(t_1)) \in \gamma, \text{ and } K^n(\tilde{\tau}(t_2)) \in h(\gamma).$$

Note that every time $\tilde{\tau}$ crosses some C', the image $\mathbf{k}(\tau)$ winds once around the cylinder $\mathbb{C}/\mathbb{Z} \simeq \hat{\mathbb{C}} \backslash \{0, \infty\}$. Hence, we have the following two possibilities:

1. $\tilde{\tau}$ crosses finitely many connected components of $\cup_{n \in \mathbb{Z}} K^n(C)$;
2. $\lim_{t \to 1-} \mathbf{k}(\tau(t)) \in \{0, \infty\}$.

In case (2), we are done.

Standard considerations of dynamics on $J(K)$ imply that for every $x \in \mathbb{R}_{\geq 0}$ which is not a preimage of 0, there exists an infinite sequence of cross-cuts $l_k \equiv K^{-n_k}(\gamma_i)$ for some choice of the inverse branch and $i = 1, 2$, such that denoting N_k the cross-cut neighborhood of l_k in $\hat{\mathbb{C}} \backslash \mathbb{R}_{\geq 0}$, we have

$$\overline{N_k} \cap \mathbb{R} \ni \{x\}.$$

In view of (1) and (2) this implies that x cannot be accumulated by $\tilde{\tau}$. Since points x as above form a dense set on $\mathbb{R}_{\geq 0}$, we see that there exists a limit

$$s = \lim_{t \to 1-} \tilde{\tau}(t) \text{ and } s \in \cup K^{-n}(0).$$

In view of (1), this implies that

$$\lim_{t \to 1-} \mathbf{k}(\tau(t)) \in \{0, \infty\}.$$

3.4 Parabolic Renormalization of f_0

We now have

Proposition 3.22 *The maps f_0 and K are both renormalizable.*

Proof The arguments are identical, so we only argue for f_0. Let us replace $\mathbf{h}_{f_0}$ by ch_{f_0} if needed so that $\mathbf{h}'_{f_0}(0) = 1$. We have to show that the coefficient a in the Taylor expansion

$$\mathbf{h}_{f_0}(z) = z + az^2 + \cdots$$

is not equal to 0, or in other words, that $z = 0$ is a simple parabolic point of $\mathbf{h}_{f_0}$. Let P_0 be an attracting petal of $z = 0$. For $n \geq 1$, denote P_{-n} the component of the preimage $(\mathbf{h}_{f_0})^{-1}(P_{-(n-1)})$ that contains $P_{-(n-1)}$. We claim that there exists n such that P_{-n} contains the unique critical value v of $\mathbf{h}_{f_0}$. Assume the contrary. Then for every n, the domain P_{-n} is a topological disk. Denote

$$B_0 = \cup P_{-n} \subset \hat{W}^+.$$

The Fatou coordinate ϕ_A of $\mathbf{h}_{f_0}$ extends to all of B_0 via the functional equation

$$\phi_A \circ \mathbf{h}_{f_0}(z) = \phi_A(z) + 1$$

as an unbranched analytic map. It is trivial to see that its image is the whole complex plane $\mathbb{C}$, and hence B_0 must be a parabolic domain. This is impossible, however, as B_0 is a subset of a Jordan domain $\hat{W}^+ \subset \mathbb{C}$.

Now assume that $a = 0$. Then the parabolic point $z = 0$ has at least two distinct attracting directions $v_1, v_2 \in S^1$. Consider corresponding attracting petals $P_A^1 \cap P_A^2 = \emptyset$. As we have shown above, the critical value v must be contained in the intersection of their preimages. This leads to a contradiction, as in this case

$$\lim \mathbf{h}_{f_0}^n(v)/|\mathbf{h}_{f_0}^n(v)| = v_1 = v_2. \qquad \square$$

Recall that the parabolic renormalization

$$F \equiv \mathscr{P}(f_0) : \hat{W}^+ \to \hat{\mathbb{C}}$$

is the unique rescaling of $\mathbf{h}_{f_0}$ whose Taylor expansion at the origin has the form $F(z) = z + z^2 + \cdots$. We now turn to discussing the dynamics of F.

Definition 3.3 The points $z \in \hat{W}^+$ whose orbits do not escape $\hat{W}^+$ form the *filled Julia set* $K(F)$. Its boundary $J(F)$ is the *Julia set*.

Lavaurs [Lav] has shown the following.

Theorem 3.4 ([Lav]) *The interior of the filled Julia set $K(F)$ coincides with the basin of the parabolic point 0. Repelling periodic orbits of F are contained in $J(F)$ and are dense in $J(F)$.*

Let $P^0 \equiv P_A$ be an attracting petal of F. For $n \in \mathbb{N}$ inductively define P^{-n} to be the connected component of the inverse image $F^{-1}(P^{-(n-1)})$ which contains $P^{-(n-1)}$. The same considerations as in the proof of Proposition 3.22 imply:

Proposition 3.23 *There exists $n \in \mathbb{N}$ such that P^{-n} is a topological disk which contains v in its closure.*

3.4.1 A Note on the General Theory for Analytic Maps of Finite Type

As the previous example clearly demonstrates, when f has global covering properties, we can expect the Écale-Voronin map $\mathcal{E}_f$ to also possess a well-understood global covering structure. The appropriate setting for a global structure theory for Écalle-Voronin maps is that of analytic maps *of finite type*, developed in A. Epstein's thesis [Ep]. While we will not need these results in our investigation, we will briefly mention some of them below for the sake of completeness of the exposition.

Definition 3.4 ([Ep]) Let $f : W \to X$ be an analytic map between two Riemann surfaces. Assume further that X is compact and W lies in some compact surface Y. Suppose that f is nowhere constant, and that every isolated singularity of f is essential. We say that f is a map of finite type if $\mathrm{Sing}(f)$ is a finite set.

Rational endomorphisms $f : \hat{\mathbb{C}} \to \hat{\mathbb{C}}$ are obviously maps of finite type. Epstein demonstrated:

Theorem 3.5 ([Ep]) *If an analytic map f with a parabolic cycle is of finite type, then the corresponding Écalle-Voronin maps $\mathcal{E}_f : W_f \to \hat{\mathbb{C}}$ inherit the same property.*

Epstein further showed that the familiar properties, such as the density of repelling periodic points and topological minimality, hold for Julia sets of analytic dynamical systems of finite type. Most importantly, he proved that the Fatou-Sullivan structure theory holds for Fatou sets of such dynamical systems:

Theorem 3.6 ([Ep]) *Every connected component of the Fatou set of a map of finite type is pre-periodic. All periodic Fatou components are either basins (attracting or parabolic), or rotation domains (Siegel disks or Herman rings).*

3.5 A Class of Analytic Mappings Invariant Under $\mathscr{P}$

3.5.1 Definition of $\mathbf{P_0}$

We will now specialize to a much narrower class of analytic maps with parabolic orbits. As before, we write

$$f_0(z) = z + z^2,$$

and we set

$$F \equiv \mathscr{P}(f_0) \text{ and } W_F = \mathscr{D}(\mathscr{P}(f_0)).$$

Definition 3.5 We will let $\widetilde{\mathbf{P}}$ be the class of analytic germs $f(z)$ at the origin which have maximal analytic extensions

$$f : \mathscr{D}(f) \to \hat{\mathbb{C}}$$

such that

(I) $\mathscr{D}(f)$ is a simply-connected domain;
(II) denoting by φ_f the Riemann mapping

$$\varphi_f : \mathbb{D} \to \mathscr{D}(f), \quad \text{with } \varphi_f(0) = 0 \text{ and } \varphi'_f(0) > 0,$$

we have

$$f \circ \varphi_f(z) = v \cdot F \circ \varphi_F(e^{2\pi i \theta} z), \quad \text{for } v \neq 0, \quad \theta \in \mathbb{R}.$$

Note that denoting by c and c^F the unique critical values of f and $F = \mathscr{P}(f_0)$ respectively, we get the rescaling factor as $v = c/c^F$.

If we again let $\mathbf{k}$ be the normalized $\mathbf{h}_K$ for the Koebe function $K(z)$, then, by Proposition 3.17, the property (II) in the above definition is equivalent to the following:

(II') denoting by φ_f the Riemann mapping

$$\varphi_f : \mathbb{D} \to \mathscr{D}(f), \quad \text{with } \varphi_f(0) = 0 \text{ and } \varphi'_f(0) > 0,$$

we have

$$f \circ \varphi_f(z) = v \cdot \mathbf{k}(e^{2\pi i \theta} z), \quad \text{for } v \neq 0, \quad \theta \in \mathbb{R}.$$

We note:

Proposition 3.24 *For all $f \in \widetilde{\mathbf{P}}$ we have $f''(0) \neq 0$.*

Proof Suppose $f''(0) = 0$. By multiplying by a nonzero constant, if necessary, we may reduce the proof to the case when $f'(0) = 1$, so that $f(z) = z + az^n + \cdots$ for $n > 2$. A Leau-Fatou flower of f has $n - 1 \geq 2$ attracting petals. Since f has a single critical value c^f, there exists an ample petal P_A such that

$$P_A \cap \bigcup_{k \in \mathbb{N}} f^k(c^f) = \emptyset.$$

Let ϕ_A be an attracting Fatou coordinate defined on P_A. Set $P_0 \equiv P_A$ and inductively define P_{-n} as the component of the preimage $f^{-1}(P_{-(n-1)})$ which contains $P_{-(n-1)}$ for $n \in \mathbb{N}$. The function ϕ_A analytically extends via the functional equation

$$\phi_A \circ f(z) = \phi_A(z) + 1$$

to the union

$$U \equiv \cup P_{-n} \subset \mathscr{D}(f).$$

A simple induction shows that ϕ_A is univalent on P_{-n}, and hence on all of U. On the other hand, $\phi_A(U) = \mathbb{C}$, which is impossible since U is a hyperbolic domain. $\Box$

Definition 3.6 We define $\widehat{\mathbf{P}}$ as the set of maps $f \in \widetilde{\mathbf{P}}$ of the form

$$f(z) = z + z^2 + \cdots$$

at the origin.

We further define $\mathbf{P} \subset \widehat{\mathbf{P}}$ as the collection of maps $f \in \widehat{\mathbf{P}}$ such that the domain $\mathscr{D}(f)$ has locally connected boundary. Finally, $\mathbf{P}_0 \subset \mathbf{P}$ consists of maps whose domain of definition is Jordan.

Since $F = \mathscr{P}(f_0) \in \mathbf{P}_0$, we have

Proposition 3.25 *The class $\mathbf{P}_0$ is nonempty.*

We now set out to prove the following theorem:

Theorem 3.7 *Every $f \in \mathbf{P}_0$ is renormalizable, and the parabolic renormalization $\mathscr{P}(f) \in \mathbf{P}_0$.*

Let $f \in \widetilde{\mathbf{P}}$ and let P_0 be an attracting petal of the parabolic point $z = 0$. For $n \geq 1$ let P_{-n} be the component of $f^{-1}(P_{-(n-1)})$ which contains $P_{-(n-1)}$. We let

$$B_0^f \equiv \cup P_{-n},$$

and call it the *immediate basin* of 0. We note:

Proposition 3.26 *For $f \in \widetilde{\mathbf{P}}$ the unique critical value $c^f \in B_0^f$. Moreover, there exists an attracting petal $P_A \subset B_0^f$ which is a topological disk containing c^f in its interior.*

Proof Assume that $v \notin B_0^f$. The attracting Fatou coordinate ϕ_A^f extends from P_0 to all of B_0^f via the functional equation

$$\phi_A^f \circ f(z) = \phi_A^f(z) + 1.$$

A simple induction shows that P_{-n} is an increasing sequence of topological disks, on each of which ϕ_A is univalent. Hence, ϕ_A restricted to B_0^f is a conformal homeomorphism onto the image. Yet it is clear that the image of ϕ_A restricted to B_0^f is the whole of $\mathbb{C}$, which contradicts the fact that B_0^f is a hyperbolic domain. The second part of the statement is elementary, and is left to the reader. $\Box$

Pushing the argument a little further, we have:

Proposition 3.27 *For $f \in \tilde{\mathbf{P}}$ the immediate basin B_0^f is simply connected and contains exactly one critical point of f. The restriction $f : B_0^f \to B_0^f$ is a degree-2 branched covering.*

Proof By Proposition 3.26, there exists n such that P_{-n} contains the critical value c^f. Inductively applying Lemma 3.1, we see that for $k \in \mathbb{N}$, the domain $P_{-(n+k)}$ is a topological disk, and $f : P_{-(n+k)} \to P_{-(n+k-1)}$ is a branched covering of degree 2 with a single, simple critical point. $\square$

The proof of Theorem 3.7 will rely on the following key result, which is a direct analogue of Theorem 3.2.

Theorem 3.8 *Let $f \in \mathbf{P}_0$, and denote B_f^0 the immediate basin of the parabolic point 0 of f. Then B_f^0 is a Jordan domain. Denote $\hat{f}$ the continuous extension of f to ∂B_0^f. There exists a homeomorphism $\rho : \partial B_0^f \to \mathbb{T}$ such that*

$$\rho(\hat{f}(z)) = 2\rho(z) \bmod 1.$$

The proof of Theorem 3.8 is quite involved, as it will require a detailed understanding of the covering properties of a map in $\mathbf{P}_0$. Let us show how Theorem 3.8 implies Theorem 3.7:

Proof Consider a Riemann map

$$v : \mathbb{D} \to B_0^f$$

which maps 0 to the sole critical point of f inside B_0^f. By Theorem 3.8 and Carathéodory Theorem, the Riemann map has a continuous extension to a homeomorphism $\tilde{v} : \overline{\mathbb{D}} \to \overline{B_0^f}$.

Let us further normalize the Riemann map $\mathbb{D} \to B_0^f$ by requiring that $\tilde{v}(1) = 0$. This specifies the mapping uniquely, and we denote it by ψ_f.

By Theorem 2.9,

$$(\psi_f)^{-1} \circ f \circ \psi_f = B : \mathbb{D} \to \mathbb{D}, \quad \text{where } B(z) = \frac{3z^2 + 1}{3 + z^2}.$$

Let us set
$$\chi = \psi_{f_0} \circ (\psi_f)^{-1} : B_0^f \mapsto B_0^{f_0}.$$

By the discussion above, this mapping is a conjugacy:

$$\chi \circ f|_{B_0} = f_0 \circ \chi|_{B_0}.$$

Let us denote (h^+, h^-) the Écalle-Voronin invariants of f. By Proposition 3.16, the germ h^+ has a maximal analytic continuation to a map in $\widetilde{\mathbf{P}}$. By Proposition 3.24, the map f is renormalizable and $\mathscr{P}(f) \in \widehat{\mathbf{P}}$. Finally, by Theorem 2.8 the germ h^+ has a maximal analytic continuation to a Jordan domain $\mathscr{D}(h^+)$, and hence $\mathscr{P}(f) \in \mathbf{P}_0$. $\qquad\square$

Let $f \in \mathbf{P}_0$. By definition of parabolic renormalization, for any repelling petal P_R^f the projection of the intersection $P_R^f \cap B_0^f$ by $\mathrm{ixp} \circ \phi_R^f$ lies in the domain $\mathscr{D}(\mathscr{P}(f))$. We remark that it covers all of it:

Remark 3.1 For any choice of P_R^f, the projection

$$\mathrm{ixp} \circ \phi_R^f (P_R^f \cap B_0^f) \cup \{0, \infty\}$$

is a union of two disjoint Jordan domains $W^+ \ni 0$ and $W^- \ni \infty$ with

$$W^+ = \mathscr{D}(\mathscr{P}(f)).$$

We begin the proof with a lemma:

Lemma 3.6 *Let D be a Jordan domain containing 0 such that*

- $c^f \notin D$;
- *there exists a Jordan arc $\tau \ni c^f$ in $\widehat{\mathbb{C}} \backslash D$ running to ∞ such that $\tau \cap B_0^f$ is connected.*

Analytically extend g to D. If $w \in B_0^f \cap D$, then $g(w) \in B_0^f$.

Proof Standard considerations imply that g extends to a branch $\hat{g}$ of f^{-1} that is defined and analytic on $\mathbb{C} \backslash \tau$. Note that $B_0^f \backslash \tau$ is also connected. For a point z_0 on the negative real axis and near to 0, the asymptotic development for the attracting Fatou coordinate (Proposition 2.5) implies that z_0 and $g(z_0) \approx z_0$ both lie in B_0^f.

Let $z \in B_0^f \backslash \tau$; then there is a Jordan arc γ in $B_0^f \backslash \tau$ from z_0 to z. By the covering properties of f, there is a unique lift $\tilde{\gamma}$ of γ starting at $g(z_0)$. Furthermore, this lift is contained in B_0^f, so its endpoint is in B_0^f. On the other hand, for $\hat{g}$ as above,

$$s \mapsto \hat{g}(\gamma(s))$$

is another lift with the same starting point. By uniqueness of lifts, it coincides with $\tilde{\gamma}$. In particular, the endpoint of $\tilde{\gamma}$ is $\hat{g}(z)$. Since we already know that $\tilde{\gamma}$ lies in B_0^f, it follows that $\hat{g}(z) \in B_0^f$. In particular, if $z \in D$, then $\hat{g}(z) = g(z)$, so $g(z) \in B_0^f$, as asserted. $\qquad\square$

Proof (Remark 3.1) Extend the local inverse g to all of P_R^f. If necessary, replace P_R^f with $g^n(P_R^f)$ for a sufficiently large $n \in \mathbb{N}$ to guarantee that there exists a domain D as described in Lemma 3.6 such that $P_R^f \subset D$.

Denote

$$W = \text{ixp} \circ \phi_R^f (P_R^f \cap B_0^f) \quad \text{and} \quad V = \text{ixp} \circ \phi_R^f (P_R^f \setminus B_0^f).$$

Lemma 3.6 implies that W and V are disjoint. The common boundary $J = \partial W = \partial V$ is the projection

$$J = \text{ixp} \circ \phi_R^f (P_R^f \cap \partial B_0^f).$$

By Theorem 3.8, J is a union of two disjoint Jordan curves. Hence, W is a union of two Jordan domains $W^+ \ni 0$ and $W^- \ni \infty$, bounded by the components of J. On the other hand, $\mathscr{D}(\mathscr{P}(f))$ is also a Jordan domain, which does not intersect J. Hence, it is contained in the component of W which surrounds 0, and is, in fact, equal to it by the maximality of the analytic continuation of $\mathscr{P}(f)$. $\square$

3.6 The Structure of the Immediate Parabolic Basin of a Map in P

In this section we will prove several results about lifts of parametrized paths of the form

$$s : [0, 1] \to \hat{\mathbb{C}}.$$

We will always assume $s(t)$ to be continuous on $(0, 1)$. We will say that s *lands* at a point $a \in \hat{\mathbb{C}}$ if

$$\lim_{t \to b} s(t) = a \quad \text{for } b \in \{0, 1\}.$$

We will generally use the same letter s to denote the function $s(t)$ and the curve $s([0, 1])$ which is its range. We will call the image of the open interval $(0, 1)$ under $s(t)$ an *open* path, and will denote it by $\overset{\circ}{s}$. Several times we will encounter the situation when there is a domain $W \subset \hat{\mathbb{C}}$ and a curve

$$s : (0, 1) \mapsto W,$$

which lands at a point $w \in \partial W$ (to fix the ideas, assume that $s(0) = w$). If w has more than one prime end in W, then there is a unique prime end $\hat{w}$ such that for every prime end neighborhood $N(\hat{w})$, we have $s \cap N(\hat{w}) \neq \emptyset$. In this case, we will write

$$s(t) \underset{t \to 0+}{\longrightarrow} \hat{w}.$$

Let $f \in \mathbf{P}$. Let $s \mapsto \gamma(s)$ be a continuous path such that:

- $\gamma(0) = \gamma(1) = 0$;
- $\gamma(s) \neq c^f$ for all $s \in (0, 1)$;
- the winding number $W(\gamma, c^f) = 1$;
- there is an $\varepsilon \in (0, \pi/2)$ such that $\gamma \cap D_\varepsilon(0)$ lies in the sector

$$\{\operatorname{Arg}(z) \in (-\pi - \varepsilon, -\pi + \varepsilon)\}.$$

Standard path-lifting considerations imply that there exists a unique continuous mapping $s \mapsto \tilde{\gamma}(s)$, defined for $0 \leq s < 1$ such that

$$\tilde{\gamma}(0) = 0, \quad \text{and} \quad f(\tilde{\gamma}(s)) = \gamma(s) \quad \text{for } 0 \leq s < 1.$$

We claim:

Proposition 3.28 *We have the following.*

1. *For any loop γ as above, there exists a limit*

$$t \in \partial\mathscr{D}(f) = \lim_{s \to 1} \tilde{\gamma}(s).$$

2. *This point (which we will denote t^f) is the same for all loops γ satisfying the above properties.*
3. *For $f = F \equiv \mathscr{P}(f_0)$, the point t^F is the projection of the inverse orbit*

$$z_{-n} = (g_0)^n(-1)$$

by $\mathrm{ixp} \circ \phi_R$.
4. *Similarly, for $f = H \equiv \mathscr{P}(h)$, the point t^H is the projection of the inverse orbit*

$$z_{-n} = K^{-n}(1),$$

where the inverse branch is selected to preserve the interval $(0, 1)$.

Proof (Proposition 3.28) We will prove (1)–(3) for $f = F \equiv \mathscr{P}(f_0)$. By the definition of **P** and by the Carathéodory Theorem, this will imply (1) and (2) in the general case. The proof of (4) follows along the same lines as the proof of (3), and will be left to the reader.

Let us begin by selecting a simple curve $\sigma \subset \hat{\mathbb{C}}$ that connects 0 and ∞, does not intersect with γ except at the endpoint 0, and such that an analytic branch of the logarithm defined on a neighborhood of σ has bounded imaginary part.

Recall that the attracting Fatou coordinate $\phi_A^{f_0}$ holomorphically extends to the whole immediate basin $B_0^{f_0}$ via the equation

$$\phi_A^{f_0} \circ f_0(z) = \phi_A^{f_0}(z) + 1.$$

This extension is a branched covering $B_0^{f_0} \to \mathbb{C}$ with simple ramification points at preimages of the critical point $-1/2$.

It is elementary to see what the lift of σ to the dynamical plane of f_0 looks like. We summarize its properties below, and invite the reader to verify them. Let us denote

$$\overset{\circ}{\sigma} \equiv \sigma \backslash \{0, \infty\}.$$

The preimage of $\overset{\circ}{\sigma}$ under $\mathrm{ixp} \circ \phi_A^{f_0}$ is a countable collection of disjoint simple curves $\cup \overset{\circ}{s}_j$. We will denote the closure of $\overset{\circ}{s}_j$ by s_j. It is obtained by adjoining two endpoints to $\overset{\circ}{s}_j$: two elements of the grand orbit $\cup_{n \geq 0} (f_0)^{-n}(0)$, that are not necessarily distinct (Fig. 3.3).

The curves s_j form a grand orbit under f_0 as well. The connected components

$$S_i \quad \text{of } B_0^{f_0} \backslash \cup s_j$$

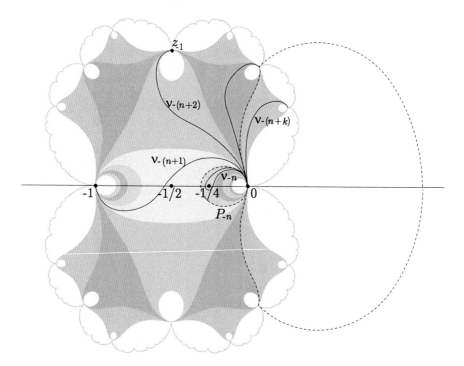

Fig. 3.3 An illustration of the proof of Proposition 3.28

are mapped by the attracting Fatou coordinate $\phi_A^{f_0}$ onto curvilinear strips $\tilde{S}_i \subset \mathbb{C}$ of infinite height and of unit width. The strips $\tilde{S}_i$ are bounded by unit translates of the same curve $\tilde{\sigma} \subset \mathbb{C}$. Let us enumerate these strips in such a way that

$$T : \tilde{S}_i \to \tilde{S}_{i+1}, \quad \text{where } T(z) = z + 1.$$

For a fixed $\tilde{S}_i$ we have

$$\sup_{x,y \in \tilde{S}_i} (\mathrm{Re}(x) - \mathrm{Re}(y)) < \infty.$$

This is where we have used the assumption on the branches of log in $\hat{\mathbb{C}} \backslash \sigma$. Thus for every $\ell \in \mathbb{Z}$, the union $\cup_{i \geq \ell} \tilde{S}_i$ contains a right half-plane.

Now let us take *any* component $S_i \subset B_0^{f_0}$. The above observation implies that for any $\ell \in \mathbb{N}$,

$$\phi_A^{f_0}(\cup_{k \geq \ell} f_0^k(S_i)) \supset \{\mathrm{Re}\, z > A\}$$

for some $A \in \mathbb{R}$. Note that if ℓ is large enough, then $\cup_{k \geq \ell} f_0^k(S_i)$ does not contain any preimages of $-1/2$, and hence the restriction of $\phi_A^{f_0}$ to it is unbranched. Denote the boundary components of $S_{i+\ell}$ by s and $f_0(s)$. Then s bounds an attracting petail P, and

$$P = \cup_{k \geq \ell} f_0^k(S_i).$$

We now complete the proof as follows. Fix

$$S \equiv S_{i+\ell} = P \backslash \overline{f_0(P)}.$$

The curve $\overset{\circ}{\gamma} = \gamma((0, 1))$ has a univalent pull-back $\overset{\circ}{\nu} \subset S$. We continuously extend it to a parametrized loop

$$\nu : [0, 1] \to S \cup \{0\}$$

which starts and ends at the parabolic point 0.

Denote $P_0 \equiv P$ and, for $j \in \mathbb{N}$, let P_{-j} be the connected component of $(f_0)^{-1}(P_{-(j-1)})$ which contains $P_{-(j-1)}$. By Proposition 3.26, there exists $n \in \mathbb{N}$ such that P_{-n} is an attracting petal such that the critical value $-1/4$ is contained in $P_{-n} \backslash \overline{P_{-(n-1)}}$. Denote $\nu_0 = \nu$ and for $1 \leq j \leq n$ let $\nu_{-j} \subset P_{-j} \backslash P_{-(j-1)}$ be the univalent pull-back of $\nu_{-(j-1)}$. The curve ν_{-n} is a parametrized loop

$$\nu_{-n} : [0, 1] \to B_0^{f_0} \quad \text{with } \nu_{-n}(0) = \nu_{-n}(1) = 0.$$

By the choice of n, the winding number

$$W(\nu_{-n}, -1/4) = 1.$$

Consider the parameterized curve

$$v_{-(n+1)} \subset P_{-(n+1)} \backslash P_{-n}$$

such that $v_{-(n+1)}(0) = 0$ and

$$f_0(v_{-(n+1)}) = v_{-n}.$$

Evidently,

$$v_{-(n+1)}(1) = -1,$$

which is the *other* preimage of the parabolic point $z = 0$ under f_0. To complete the argument, let us now consider the connected component W of

$$B_0^{f_0} \backslash \overline{P_{-(n+1)}} \supset W$$

which contains the "upper" preimage of -1, the point

$$z_{-1} \equiv \frac{-1 + \sqrt{3}i}{2} \in \mathbb{H},$$

in the boundary. Let $v_{-(n+2)} \subset W$ be the next preimage,

$$f_0(v_{-(n+2)}) = v_{-(n+1)}, \quad \text{with } v_{-(n+2)}(0) = 0 \quad \text{and} \quad v_{-(n+2)}(1) = z_{-1}.$$

The inverse branch g_0 of the quadratic map f_0 univalently extends to a map

$$g_0 : W \to W.$$

For $k \geq 3$ denote $v_{-(n+k)}$ the preimage of $v_{-(n+k-1)}$ by this branch. By the Denjoy-Wolff Theorem,

$$v_{-(n+k)} \to 0.$$

Furthermore, $v_{-(n+k)}$ is disjoint from $P_{-(n+1)}$. For any given ample repelling petal P_R there exists k such that $v_{-(n+k)} \subset P_R$. Consider the projection

$$\tilde{\gamma} \equiv \mathrm{ixp} \circ \phi_R^{f_0}(v_{-(n+k)}).$$

By construction, it satisfies the properties (1)–(3).   □

To help understand the shape of the immediate basin of $F \in \mathbf{P}_0$, let us look at the drawing in Fig. 3.4. It illustrates the case $F = \mathscr{P}(f)$ (the reader may think of $f = f_0$, to fix the ideas). The left figure shows a fragment of the boundary of the immediate basin B_0^f near the parabolic fixed point 0. If we look on the right, we see a schematic picture of the Jordan domain $\mathscr{D}(\mathscr{P}(f))$. The point $t^{\mathscr{P}(f)} \in \partial \mathscr{D}(\mathscr{P}(f))$ is the "tip of

Fig. 3.4 The formation of a "tail" in the immediate parabolic basin of $\mathscr{P}(f)$

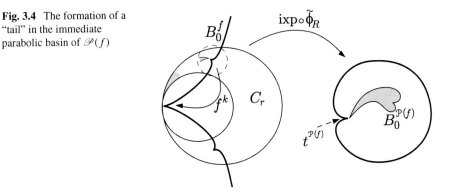

the tail" of the immediate basin $B_0^{\mathscr{P}(f)}$. On the left the reader can see how the tail is formed. The lift of the basin B_0^f fits inside a suitable repelling crescent C_R^f, reaching to the upper tip of the crescent (which under $\mathrm{ixp} \circ \phi_R^f$ becomes the parabolic point $z = 0$ on the right). The point $t^{\mathscr{P}(f)}$ lifts under $\mathrm{ixp} \circ \phi_R^f$ to an f^{-k}-preimage $w \in C_R^f$ of the parabolic point 0. The lift of the immediate basin $B_0^{\mathscr{P}(f)}$ to C_R^f reaches to w; its shape near w is a conformal image of the shape near 0 (a tail). The map $\mathrm{ixp} \circ \phi_R^f$ is conformal in a neighborhood of w, and hence the basin $B_0^{\mathscr{P}(f)}$ also has a tail ending at $t^{\mathscr{P}(f)}$.

3.6.1 Definition of $\mathbf{P}_1$ and Puzzle Partitions

Definition 3.7 We let $\mathbf{P}_1$ to be the collection of maps $f \in \mathbf{P}$ for which the point t^f is accessible from the complement of the domain $\mathscr{D}(f)$. We thus have

$$\mathbf{P}_0 \subset \mathbf{P}_1 \subset \mathbf{P}.$$

Theorem 3.8 will follow from a stronger statement:

Theorem 3.9 Let $f \in \mathbf{P}_1$, and denote B_f^0 the immediate basin of the parabolic point 0 of f. Then B_f^0 is a Jordan domain. Denote by $\hat{f}$ the continuous extension of f to ∂B_0^f. There exists a homeomorphism $\rho : \partial B_0^f \to \mathbb{T}$ such that

$$\rho(\hat{f}(z)) = 2\rho(z) \bmod 1.$$

Let us make a new definition. Let $f \in \mathbf{P}_1$. Consider an attracting petal P_A^f that contains the unique critical value $c^f \in B_0^f$. Let $\tau \subset B_0^f$ be a simple curve that connects c^f with the parabolic point 0 and has the property $f(\tau) \subset \tau$. To fix the ideas, we will take this curve to be the horizontal ray

$$\{\operatorname{Re}(z) \geq \operatorname{Re}(\phi_A^f(c^f)), \quad \operatorname{Im}(z) = \operatorname{Im}(\phi_A^f(c^f))\}$$

in the Fatou coordinate. Using Proposition 3.28, and passing from a curve around the critical value to a slit connecting the critical value with a parabolic point, we see:

Proposition 3.29 *There exists a unique simple curve Γ^{in} connecting t^f with 0 such that*

$$f(\Gamma^{in}) = \tau.$$

By definition of $\mathbf{P}_1$, the tip of the tail $t^f \in \partial \mathscr{D}(f)$ is accessible from outside $\mathscr{D}(f)$. Let us continue Γ^{in} by attaching a simple curve Γ^{out} connecting t^f to ∞ without intersecting $\overline{\mathscr{D}(f)}$. Let us further require that Γ^{out} coincides with a negative real ray in a neighborhood of ∞.

Definition 3.8 We call $\Gamma \equiv \overline{\Gamma^{in} \cup \Gamma^{out}}$ the *primary cut* of f.

We prove the following:

Proposition 3.30 *Let $f \in \mathbf{P}_1$. There exists a domain $U \subset \mathbb{C}$ such that:*

(1) *U is a Jordan domain;*
(2) *$U \subset \mathscr{D}(f)$ and $\partial U \cup \partial \mathscr{D}(f) = \{t^f\}$;*
(3) *$\partial U \subset f^{-1}(\overline{\Gamma^{out}} \cup \partial \mathscr{D}(f))$ and $U \supset \Gamma^{in}$;*
(4) *$U \ni c^f$;*
(5) *there is a single critical point $p^f \in \Gamma^{in}$ of f inside U.*
     Finally, there is a simple arc $\Gamma_{-1}^{in} \subset f^{-1}(\Gamma^{in}) \cap U$ with endpoints $u^+, u^- \in \partial U$ such that $\Gamma_{-1}^{in} \cap \Gamma^{in} = \{p^f\}$ for which the following holds:
(6) *consider the Jordan domain U' which is the connected component of $U \backslash \Gamma_{-1}^{in}$ whose boundary does not contain the point t^f. Then*

$$f : U' \backslash \Gamma^{in} \longrightarrow \mathscr{D}(f) \backslash \Gamma^{in}$$

     is a univalent map.

In view of the definition of $\mathbf{P}_1$, it suffices to prove a slightly more general statement for $f = F \equiv \mathscr{P}(f_0)$:

Proposition 3.31 *Let $F = \mathscr{P}(f_0)$. Consider any simple arc $\gamma = \gamma_1 \cup \gamma_2$ and any Jordan curve τ with the following properties:*

- *γ_1 is a simple arc which connects 0 and t^F;*
- *γ_2 is a simple arc which connects t^F and ∞;*
- *the curve γ_1 approaches 0 within a sector $\{\operatorname{Arg}(z) \in (\pi/2, 3\pi/2)\}$;*
- *the curve γ_2 approaches ∞ within a sector $\{\operatorname{Arg}(1/z) \in (\pi/2, 3\pi/2)\}$;*
- *the curve τ separates c^F from ∞ and $\tau \cap \gamma = \{t^F\}$.*

There exists a domain U such that:

1. *U is a Jordan domain;*
2. $U \subset \mathscr{D}(F)$ *and* $\partial U \cup \partial \mathscr{D}(F) = t^F$;
3. $\partial U \subset F^{-1}(\overline{\gamma_2} \cup \tau)$;
4. *U contains a single critical point* p^F;
5. *there is a simple arc* $\delta \subset U$ *which is a preimage of* γ_1, *has endpoints on the boundary of U, and intersects* γ_1 *at the critical point* p^F *only. If we let* U' *be the component of* $U \setminus \delta$ *not containing* t^F *on the boundary, then the map*

$$F : U' \longrightarrow \mathscr{D}(F) \setminus F(\delta)$$

is univalent onto the image.

Proof Let us again begin by selecting a simple curve $\sigma \subset \hat{\mathbb{C}}$ which connects 0 and ∞, and does not intersect with γ except at the point 0, and which intersects τ at a single point, such that an analytic branch of the logarithm defined on a neighborhood of σ has a bounded imaginary part.

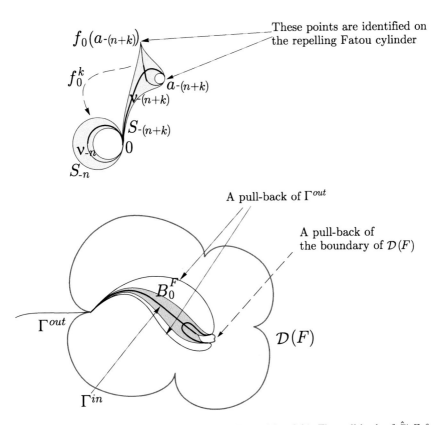

Fig. 3.5 An illustration for Proposition 3.30 and Proposition 3.31. The pull-back of $\hat{\mathbb{C}} \setminus \Gamma$ for $F = \mathscr{P}(f_0)$

Let us parametrize γ:

$$\gamma(t) : [0, 1] \to \hat{\mathbb{C}}$$

so that $\gamma(0) = 0$ and $\gamma(1) = \infty$.

With a slight abuse of notation we use the same notation for the preimages of σ and γ, etc. as in the proof of Proposition 3.28. Let S_{-n} be the fundamental crescent $P_{-n} \backslash \overline{f(P_{-n})}$ which contains the critical value $-1/4$. The corresponding component $\nu_{-n} \subset S_{-n}$ of the lift of γ is a simple arc that passes through $-1/4$, and whose two ends land at 0 (Fig. 3.5).

There are two prime ends of the point 0 in the crescent S_{-n}. We denote the "upper" one by 0^+_{-n}, and the "lower" one by 0^-_{-n}. They correspond to the points 0 and ∞ respectively in the quotient $\mathscr{C}_A \simeq \hat{\mathbb{C}}$. We write

$$\nu_{-n}(t) \xrightarrow[t \to 0+]{} 0^+_{-n} \quad \text{and} \quad \nu_{-n}(t) \xrightarrow[t \to 1-]{} 0^-_{-n}.$$

Writing $S_{-(n+k)} = P_{-(n+k)} \backslash P_{-(n+k-1)}$, we see that

$$f_0 : S_{-(n+1)} \to S_{-n}$$

is a double covering, branched at

$$-1/2 \mapsto -1/4 \in \nu_{-n}.$$

We thus obtain two parametrized curves $\nu^1_{-(n+1)}$ and $\nu^2_{-(n+1)}$ as the lifts of ν_{-n} by f_0. The intersection

$$\overset{\circ}{\nu}^1_{-(n+1)} \cap \overset{\circ}{\nu}^2_{-(n+1)} = -1/2.$$

For $j \leq n + 1$, we denote 0^+_{-j}, 0^-_{-j} the "upper" and the "lower" prime ends of 0 in S_{-j}, so that

$$\tilde{f}_0 : 0^\pm_{-j} \mapsto 0^\pm_{-(j-1)}.$$

The other preimage of 0 in $\partial S_{-(n+1)}$ is the point -1. We use $-1^+_{-(n+1)}$ and $-1^-_{-(n+1)}$ to denote its two prime ends in $S_{-(n+1)}$, labeled so that the Carathéodory extension $\tilde{f}_0$ maps

$$-1^+_{-(n+1)} \mapsto 0^+_{-n} \quad \text{and} \quad -1^-_{-(n+1)} \mapsto 0^-_{-n}.$$

We have

$$\nu^1_{-(n+1)}(t) \xrightarrow[t \to 0+]{} 0^+_{-(n+1)} \quad \text{and} \quad \nu_{-(n+1)}(t) \xrightarrow[t \to 1-]{} -1^-_{-(n+1)};$$

$$\nu^2_{-(n+1)}(t) \xrightarrow[t \to 0+]{} -1^+_{-(n+1)} \quad \text{and} \quad \nu_{-(n+1)}(t) \xrightarrow[t \to 1-]{} 0^-_{-(n+1)}.$$

Denote $S_{-(n+k)} \subset W$ the pull-back of $S_{-(n+k-1)}$ by the inverse branch g_0, and

$$v^{1,2}_{-(n+k)} \subset S_{-(n+k)}$$

the corresponding preimage of v_{-n}. A trivial induction shows that there are exactly three points

$$\{0, a_{-(n+k)}, f(a_{-(n+k)})\} \subset \partial S_{-(n+k)},$$

that map to 0 under f_0^k. Furthermore, $a_{-(n+k)}$ has two prime ends in $S_{-(n+k)}$; the points 0 and $f_0(a_{-(n+k)})$ each have a single prime end. We denote $a^{\pm}_{-(n+k)}$ the prime end mapped to $0^{\pm}_{-n}$ by the Carathéodory extension $\tilde{f}_0^k$.

We have

$$v^1_{-(n+k)}(t) \xrightarrow[t\to 0+]{} 0_{-(n+k)} \quad \text{and} \quad v_{-(n+1)}(t) \xrightarrow[t\to 1-]{} f_0(a_{-(n-k)});$$

$$v^2_{-(n+k)}(t) \xrightarrow[t\to 0+]{} a^+_{-(n+1)} \quad \text{and} \quad v_{-(n+k)}(t) \xrightarrow[t\to 1-]{} a^-_{-(n+1)}.$$

Each of the sets v_{-j} is composed of two parts, $v_{-j,1}$ and $v_{-j,2}$, which are the preimages of γ_1 and γ_2 respectively. For $j \leq n$, they intersect at a single point b_{-j}; for $j > n$ there are two such points b^1_{-j}, b^2_{-j}.

Denote $\kappa_{-(n+k)} \subset S_{-(n+k)}$ the preimage of τ. Elementary considerations of monodromy, which we spare the reader, imply that $\kappa_{-(n+k)}$ is a union of two simple arcs: one connecting 0 with $b^1_{-(n+k)}$, the other with $b^2_{-(n+k)}$, and otherwise disjoint from $v_{-(n+k)}$. We denote them as $\kappa^1_{-(n+k)}$ and $\kappa^2_{-(n+k)}$ respectively.

Considerations of the Denjoy-Wolff Theorem imply that for every ample repelling petal P_R, there exists k such that $S_{-(n+k)} \subset P_R$. Let us project the picture in $S_{-(n+k)}$ back to $\hat{\mathbb{C}}$ using $\text{ixp} \circ \phi^{f_0}_R$. The points $a_{-(n+k)}$ and $f_0(a_{-(n+k)})$ are identified in the projection. We obtain a Jordan domain U, bounded by the projections of $\kappa^1_{-(n+k)}$, $\kappa^2_{-(n+k)}$, $v_{-(n+k),2}$. It satisfies the properties (1)–(5) by the construction.   □

Corollary 3.3 *Let $f \in \mathbf{P}_1$, and let U be the domain constructed in Proposition 3.30. We have*

- $B_0^f \subset U$;
- $\partial B_0^f \cap \partial \mathscr{D}(f) = \{t^f\}$.

Proof Since B_0^f does not intersect with $\overline{\Gamma^{\text{out}}} \cup \partial \mathscr{D}(f)$ and $\partial U \subset f^{-1}(\overline{\Gamma^{\text{out}}} \cup \partial \mathscr{D}(f))$, we have $B_0^f \subset U$. Hence,

$$\partial B_0^f \cap \partial \mathscr{D}(f) \subset \{t^f\}.$$

On the other hand, the cut $\Gamma^{\text{in}} \subset B_0^f$, hence

$$t^f \in \overline{\Gamma^{\text{in}}} \in \partial B_0^f,$$

which completes the argument.   □

3.6.2 The Immediate Basin of a Map in $\mathbf{P}_1$ Is a Jordan Domain

We are now in a position to begin the proof of Theorem 3.9, which in turn implies Theorem 3.8. Let us fix $F \in \mathbf{P}_1$ and let U be the domain constructed in Proposition 3.30. The preimage $\overline{F^{-1}(\Gamma^{\text{in}})} \cap B_0^F$ consists of the curve Γ^{in} and a simple curve Γ_{-1}^{in}; the two curves cross at the critical point $p^F \in B_0^F$. To fix the ideas, let us parametrize

$$\Gamma^{\text{in}} : [0, 1] \rightarrow \overline{B_0^f},$$

so that $\Gamma^{\text{in}}(0) = 0$ and $\Gamma^{\text{in}}(1) = t^f$. Let $s_1 < s_2 \in (0, 1)$ be such that

$$\Gamma^{\text{in}}(s_1) = c^F, \quad \Gamma^{\text{in}}(s_2) = p^F.$$

By elementary path-lifting considerations, the curve Γ_{-1}^{in} is a cross-cut in U. Let V be the connected component of $U \backslash \Gamma_{-1}^{\text{in}}$ which does not contain t^F in its boundary. By construction (Proposition 3.30), there exists a univalent branch G of F^{-1} that maps

$$\mathscr{D}(F) \backslash \Gamma^{\text{in}}([s_1, 1]) \mapsto V \backslash \Gamma^{\text{in}}([s_1, s_2]).$$

We note:

Proposition 3.32 *The inverse branch G maps the domain $\mathscr{D}(F) \backslash \Gamma^{\text{in}}$ inside itself.*

Let $l(t)$ be a simple path such that $l(0) = u$, $l(1) = F(u) \in \partial \mathscr{D}(F)$, and $\overset{\circ}{l} \cap \overline{U} = \emptyset$. Since $\mathscr{D}(F)$ is a Jordan domain, the inverse branch G has a continuous extension to the boundary point u. Set $u_0 \equiv u$, and denote by u_{-n} the orbit of u_0 under G. The curve l has a univalent pull-back $l_{-1} \subset V$ such that $l_{-1}(0) = u_{-1}$ and $l_{-1}(1) = u$. We let

$$l_{-n} = G(l_{-(n-1)}) \quad \text{for } n \geq 2.$$

Set

$$\Lambda = \cup_{n \geq 1} l_{-n}.$$

We parametrize this curve by $(0, 1]$ so that

$$\Lambda \left(\left[\frac{1}{j}, \frac{1}{j-1} \right] \right) = l_{-(j-1)} \quad \text{for } j \geq 2.$$

Thus, $\Lambda(1/j) = u_{-j}$. By Proposition 3.32 and the Denjoy-Wolff Theorem we have

Proposition 3.33 *The curve Λ is disjoint from B_0^F, and lands at 0:*

$$\lim_{t \to 0-} \Lambda(t) = 0.$$

Definition 3.9 We call the curve $\hat{\Gamma} \equiv \Gamma \cup \Lambda$ the *secondary* cut.

Let $u^+, u^- \in \partial U$ be the two endpoints of Γ_{-1}^{in}, labeled in such a way that u^+ is encountered first, when going around ∂U in the positive direction from t^f. Denote the connected components of $U \backslash \hat{\Gamma}$ by U^+, U^-, so that $U^{\pm} \ni u^{\pm}$.
Consider the inverse branch of f which maps $B_0^F \backslash \Gamma^{\text{in}}([0, s_1])$ into $B_0^F \cap U^+$, and let G_0 be its univalent extension to

$$G_0 : U \backslash (\Lambda \cup \Gamma^{\text{in}}([0, s_1])) \hookrightarrow U^+.$$

Replacing U^+ with U^- we similarly define an analytic branch of F^{-1}

$$G_1 : U \backslash (\Lambda \cup \Gamma^{\text{in}}([0, s_1])) \hookrightarrow U^-.$$

Let $P \subset B_0^F$ be an attracting petal that contains c^F in its interior, and let P_{-1} be a connected component of $F^{-1}(P)$ such that

$$P \subset P_{-1}.$$

As shown in (Fig. 3.6), We set

$$\mathbf{A} \equiv U \backslash P_{-1}.$$

Let

$$\mathbf{A}_0 \equiv \mathbf{A} \cap U^+ \quad \text{and} \quad \mathbf{A}_1 \equiv \mathbf{A} \cap U^-.$$

If $i_0 i_1 \ldots i_n$ is an arbitrary sequence of $n + 1$ 0's and 1's, we define

$$\mathbf{A}_{i_0 i_1 \ldots i_n} := \{z \in \mathbf{A}_0 : f^j(z) \in \mathbf{A}_{i_j} \quad \text{for } 1 \leq j \leq n\}.$$

The proof of Theorem 3.9 now follows the same lines as in Sect. 3.2.1. It follows from the definitions that

- $\mathbf{A}_{i_0 i_1 \ldots i_n}$ is decreasing in n: $\mathbf{A}_{i_0 \ldots i_n} \subset \mathbf{A}_{i_0 \ldots i_{n-1}}$;
- $F \mathbf{A}_{i_0 i_1 \ldots i_n} = \mathbf{A}_{i_1 \ldots i_n}$;
- $\mathbf{A}_{i_0 i_1 \ldots i_n} = G_{i_0} \mathbf{A}_{i_1 \ldots i_n}$, and hence

$$\mathbf{A}_{i_0 i_1 \ldots i_n} = G_{i_0} \circ G_{i_1} \circ \cdots \circ G_{i_{n-1}} \mathbf{A}_{i_n},$$

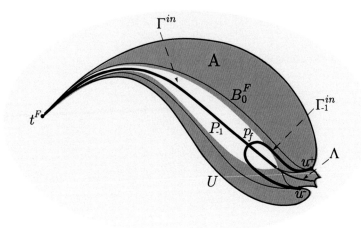

Fig. 3.6 The domain **A**

or, more generally,

$$\mathbf{A}_{i_0 i_1 \ldots i_n} = G_{i_0} \circ G_{i_1} \circ \cdots \circ G_{i_{j-1}} \mathbf{A}_{i_j i_{j+1} \ldots i_n} \quad \text{for } 1 \le j \le n-1.$$

We show that the diameters of the puzzle pieces $\mathbf{A}_{i_0 i_1 \ldots i_n}$ go to zero as $n \to \infty$, uniformly in $i_0 i_1 \ldots i_n$. More precisely, let us define

$$\rho_n := \sup \left\{ \text{diam}(\mathbf{A}_{i_0 \ldots i_n}) : i_0 \ldots i_n \in \{0, 1\}^{n+1} \right\}.$$

Since $\mathbf{A}_{i_0 \ldots i_n} \supset \mathbf{A}_{i_0 \ldots i_{n+1}}$, the sequence ρ_n is non-increasing in n, so

$$\rho_* := \lim_{n \to \infty} \rho_n \tag{3.13}$$

exists.

We will prove

Proposition 3.34 *The limit $\rho_* = 0$.*

In the same way as Lemma 3.2, we have

Lemma 3.7 *There is an infinite sequence $i_0 i_1 \ldots i_n \ldots$ so that*

$$\text{diam}(A_{i_0 i_1 \ldots i_n}) \ge \rho_* \quad \text{for all } n.$$

Proof (Proposition 3.34) Let us assume the contrary: $\rho_* > 0$.

We fix a sequence $i_0 i_1 \ldots$ as in the Lemma 3.2, and we split the proof into three cases:

1. i_j is eventually 0;
2. i_j is eventually 1;
3. neither of the above holds.

We start with case (3). There are then infinitely many j's so that

$$i_j = 0 \quad \text{and} \quad i_{j+1} = 1.$$

Let j_k be a strictly increasing sequence of such j's. Then, for each k,

$$f_{-j_k} := G_{i_0} \circ G_{i_1} \circ \cdots \circ G_{i_{j_k-1}}$$

is an analytic branch of the inverse of $f_0^{j_k}$ mapping $\mathbf{A}_{01}$ bijectively to $\mathbf{A}_{i_0\ldots i_{j_k+1}}$. The closure of $\mathbf{A}_{01}$ does not intersect the postcritical set of f. We now use a version of the argument we gave in the proof of Proposition 3.11.

We let Q be a simply connected open neighborhood of $\overline{\mathbf{A}_{01}}$ disjoint from the postcritical set. Then each f_{-j_k} extends to an analytic branch of the inverse of F^{j_k} defined on Q (and we denote this extension also by f_{-j_k}).

By Montel's Theorem, there is a subsequence of (j_k) along which f_{-j_k} converges uniformly on compact subsets of U in the spherical metric on $\hat{\mathbb{C}}$. By adjusting the notation, we can assume that the sequence (f_{-j_k}) itself converges to an analytic function which we denote by h. Since

$$\mathrm{diam}(f_{-j_k}\mathbf{A}_{01}) = \mathrm{diam}(\mathbf{A}_{i_0\ldots i_{j_k+1}})$$

does not go to zero as $k \to \infty$, the function h is nonconstant.

Let $z_0 := G_0(u^-) \in \mathbf{A}_{01} \cap \partial B_0^f$. By invarince of the basin boundary, we have $w_0 = h(z_0) \in \partial B_0^f$. Since h is nonconstant, $h(\mathbf{A}_{01})$ contains an open neighborhood W of w_0. Let $V \Subset W$ be a smaller open neighborhood of w_0. Then, an arbitrary large iterate F^{j_k} maps V inside Q, which is impossible, since $f(u^-) = t^f \in \partial \mathcal{D}(f)$.

We turn next to case 1 above, and deal first with the situation $i_j = 0$ for all j. We write

$$\mathbf{A}^{(n)} := \mathbf{A}_{\underbrace{0\ldots 0}_{n\ \text{terms}}} = (G_0)^{n-2}\mathbf{A}_{00}.$$

Now G_0 maps $\mathbf{A}_{00}$ into itself, and $G_0 = G$ on $\mathbf{A}_{00}$, so we can write

$$\mathbf{A}^{(n)} = (G)^{n-2}\mathbf{A}_{00}.$$

By the Denjoy-Wolff Theorem applied to g and local dynamics near the parabolic point 0,

$$(G)^{n-2} \to 0 \quad \text{uniformly on } \mathbf{A}_{00},$$

so

$$\mathrm{diam}(\mathbf{A}^{(n)}) \to 0 \quad \text{as } n \to \infty.$$

Next consider sequences of the form $i_0 i_1 \ldots i_k 00 \ldots$, and use the formula

$$A_{i_0 \ldots i_k \underbrace{0 \ldots 0}_{n \text{ terms}}} = G_{i_0} \circ G_{i_1} \circ \cdots \circ G_{i_k} A^{(n)}.$$

The mapping $G_{i_0} \circ G_{i_1} \circ \cdots \circ G_{i_k}$ extends to be continuous on $\overline{A_{00}}$, and $\text{diam}(A^{(n)}) \to 0$ by what we just proved, so

$$\text{diam}(A_{i_0 \ldots i_k \underbrace{0 \ldots 0}_{n \text{ terms}}}) \to 0 \quad \text{as } n \to \infty.$$

A similar argument, using $G_1 = G$ on A_{11} shows that

$$\text{diam}(A_{i_0 \ldots i_k \underbrace{1 \ldots 1}_{n \text{ terms}}}) \to 0 \quad \text{as } n \to \infty.$$

Thus, in all three cases, $\text{diam} A_{i_0 \ldots i_n} \to 0$ as $n \to \infty$, contradicting the construction of $i_0 i_1 \cdots$, so the proposition is proved. $\qquad\square$

Let $\mathbf{i} = (i_j)_{j=0}^{N}$ be a finite or infinite sequence ($N \le \infty$) of 0's and 1's. We interpret it as a binary representation of a number in $[0, 1]$:

$$\underline{\mathbf{i}}_2 \equiv \sum_{j=0}^{N} i_j 2^j \in [0, 1].$$

For any such dyadic sequence,

$$\overline{A_{i_0}} \supset \overline{A_{i_0 i_1}} \supset \cdots \supset \overline{A_{i_0 \ldots i_n}} \supset \ldots$$

is a nested sequence of compact sets in $\mathbb{C}$ with diameter going to 0, so its intersection contains exactly one point, which we denote by $\hat{z}(\mathbf{i})$. It is immediate from the construction that $\mathbf{i} \mapsto \hat{z}(\mathbf{i})$ is continuous from $\{0, 1\}^{\mathbb{Z}}$ to $\mathbb{C}$. It is not injective, however, this ambiguity is easily tractable:

Lemma 3.8 *Suppose* $\mathbf{i} = i_0 i_1 \ldots i_{n-1}$ *and* $\mathbf{i}' = i_0' i_1' \ldots i_{n-1}'$ *are two finite dyadic sequences of an equal length, and*

$$\overline{A_{\mathbf{i}}} \cap \overline{A_{\mathbf{i}'}} \ne \emptyset.$$

Then either $\underline{\mathbf{i}}_2 = \underline{\mathbf{i}}_2'$, *or* $\underline{\mathbf{i}}_2 = \underline{\mathbf{i}}_2' \pm 2^{-n}$.

The proof is a straightforward induction in n, and is left to the reader.

As a corollary, we get:

Corollary 3.4 *Let* $\mathbf{i}$ *and* $\mathbf{i}'$ *be two infinite dyadic sequences. If* $\underline{\mathbf{i}}_2 \ne \underline{\mathbf{i}}_2'$, *then* $\overline{A_{i_0 \ldots i_n}}$ *and* $\overline{A_{i_0' \ldots i_n'}}$ *are disjoint for large enough* n.

Proof For any $\mathbf{i} = i_0 \ldots i_n \ldots$, and any n

$$|\underline{\mathbf{i}}_2 - \underline{i_0 \ldots i_{n-1}}_2| \le 2^{-n},$$

so, if n is large enough so that

$$|\underline{\mathbf{i}}_2 - \underline{\mathbf{i}}'_2| > 3 \times 2^{-n},$$

then

$$|\underline{i_0 \ldots i_{n-1}}_2 - \underline{i'_0 \ldots i'_{n-1}}_2| > 2^{-n},$$

which by Lemma 3.3 implies that $\overline{\mathbf{A}_{i_0 \cdots i_{n-1}}}$ and $\overline{\mathbf{A}_{i'_0 \cdots i'_{n-1}}}$ are disjoint, as asserted.   □

Putting together what we know about the map $\mathbf{i} \mapsto \hat{z}(\mathbf{i})$, we finally obtain:

Proposition 3.35 *The angles $\mathbf{i}_2 = \mathbf{i}'_2$ if and only if $\hat{z}(\mathbf{i}) = \hat{z}(\mathbf{i}')$. Hence there is a function $\theta \mapsto z(\theta)$ from the circle $\mathbb{R}/\mathbb{Z}$ to $\mathbb{C}$ so that*

$$\hat{z}(\mathbf{i}) = \underline{\mathbf{i}}_2.$$

The function $z(\,.\,)$ is continuous, bijective, and maps the circle onto

$$J \equiv \overline{\cap_n((F^{-n})|_U \overline{\mathbf{A}})}.$$

We thus have

Corollary 3.5 *The set J is a Jordan curve.*

Proposition 3.36 *Let $\tilde{F} : J \to J$ be defined as $\tilde{F}(t^F) = 0$, and coincide with F elsewhere on J. The map $\underline{\mathbf{i}}_2 \mapsto \hat{z}(\mathbf{i})$ is a conjugacy between $\theta \mapsto 2\theta$ and $\tilde{F}(z)$:*

$$\tilde{F}(\hat{z}(\mathbf{i})) = \hat{z}(2 \cdot \underline{\mathbf{i}}_2).$$

Let us denote B the connected component of $\hat{\mathbb{C}} \backslash J$ which contains c^F. By construction, $B \cap B_0^F \ne \emptyset$. By the Maximum Principle,

$$F : B \to B.$$

By the classification of dynamics on a hyperbolic domain,

$$B \subset B_0^F.$$

On the other hand, preimages of t^f are dense in $J = \partial B$, and hence

$$B_0^F \cap J = \emptyset.$$

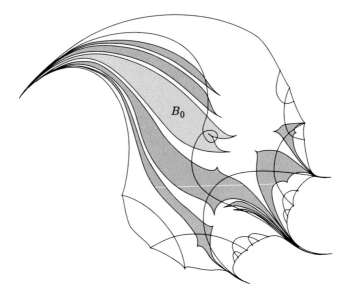

Fig. 3.7 Several components of the parabolic basin of $f \in \mathbf{P}_0$

We conclude:

$$B = B_0^F,$$

and thus Theorem 3.9 is proven.

We illustrate the topological structure of the parabolic basin of a map $f \in \mathscr{P}(\mathbf{P}_0)$ in Fig. 3.7. We have indicated the immediate basin B_0 as well as several components of $f^{-1}(B_0)$. To better understand those, we have drawn the preimage $f^{-1}(C)$ of the circle

$$C = \{|z| = |c|\},$$

which cuts across the only critical value c of f. These critical level curves partition the domain W into univalent preimages of the disks $D^+ = D_{|c|}(0)$ and $D^- = \hat{\mathbb{C}} \backslash \overline{D^+}$.

Two smooth branches in the preimage $f^{-1}(C)$ may intersect at a ramification point of f (all of the ramification points are simple). Each *critical component* of $f^{-1}(B_0)$ has one of these branch points, and touches ∂W in two points ("preimages" of the asymptotic value 0). The *non-critical components* of $f^{-1}(B_0)$ arrange themselves in sequences along the critical level curves, as shown.

3.7 Convergence of Parabolic Renormalization

By definition of the class $\mathbf{P}_0$, every map $f \in \mathbf{P}_0$ can be decomposed as

$$f = v \cdot \mathscr{P}(f_0) \circ \varphi_f(e^{2\pi i \theta} z) : \mathscr{D}(f) \to \hat{\mathbb{C}},$$

where $v = c^f/c^{\mathscr{P}(f_0)}$. We will denote $\psi_f = \varphi_f^{-1}$. Thus, φ_f conformally maps the unit disk $\mathbb{D}$ onto $W_f = \text{Dom}(f)$ with $\psi_f(0) = 0$ and $\psi_f'(0) > 0$. We will topologize $\mathbf{P}_0$ by identifying it with the space of thus normalized conformal maps of the unit disk:

$$f \mapsto \psi_f,$$

equipped with the compact-open topology.

Let us state an obvious consequence of the Koebe Distortion and Ascoli-Arzelà Theorems:

Lemma 3.9 *Let $\mathscr{S}$ be a family of univalent maps $h : \mathbb{D} \to \mathbb{C}$ with $h(0) = 0$. Assume further that there exist positive constants $0 < a < b$ such that $a \leq |h'(0)| \leq b$ for every $h \in \mathscr{S}$. Then the family $\mathscr{S}$ is equicontinuous on compact subsets of $\mathbb{D}$.*

Suppose further that there exists $M > 0$ such that $|h(0)| < M$ for all $h \in \mathscr{S}$. Then the family $\mathscr{S}$ is precompact in the sense of locally uniform convergence.

In what follows it will be useful for us to distinguish between analytic *maps* defined in a neighborhood of the origin, and *germs* of analytic maps at the origin, that is, elements of $\mathbb{C}\{z\}$. A map f will always have a specified domain (not necessarily natural in any sense). The germ of f at $z = 0$ will be denoted $\hat{f}$. Let us denote by $\mathbf{W}$ the class of maps $f : \Omega_f \to \mathbb{C}$ where $\hat{f}$ is a simple parabolic germ at the origin of the form $\hat{f}(z) = z + z^2 + \cdots$ and $\Omega_f \ni 0$ is a connected subdomain of $\mathbb{C}$ such that:

1. $f : (\Omega_f \backslash \{0\}) \to \mathbb{C}^*$ is a branched covering;
2. $f|_{\Omega_f}$ has a single critical value and all critical points of f are simple.

Note that in the definition of a map $f : \Omega_f \to \mathbb{C}$ in $\mathbf{W}$ we do not assume any maximality of the choice of the domain Ω_f. In particular, the same map can have two different restrictions satisfying the above conditions that correspond to two different points in $\mathbf{W}$.

The celebrated result of H. Inou and M. Shishikura [IS] can be stated as follows.

Theorem 3.10 ([IS]) *There exists a class $\mathbf{F}$ of analytic maps*

$$f : D_f \to \mathbb{C}$$

(with a specified domain), with a simple parabolic fixed point at the origin, of the form $\hat{f}(z) = z + z^2 + \cdots$ such that the following properties hold:

- *for every $f : \Omega_f \to \mathbb{C}$ in $\mathbf{W}$ with the exception of $f(z) = z + z^2$, the map f has a restriction to a domain D_f which belongs to $\mathbf{F}$;*
- *the domain D_f contains a simple critical point v_f of f, whose image c_f also lies in D_f;*
- *every $f \in \mathbf{F}$ is renormalizable and $\mathscr{P}(\mathbf{F}) \subset (\mathbf{F})$;*
- *$\mathscr{P}(\mathbf{F})$ is compact in the sense of compact-open topology;*
- *there exists a map $f_* \in \mathbf{F}$ which is a fixed point of the parabolic renormalization:*

$$\mathscr{P}(f_*) = f_*;$$

- *for $f_0(z) = z + z^2$ there is a domain D_0 such that $\mathscr{P}(f_0)$ analytically continues to D_0 and*

$$\mathscr{P}(f_0)|_{D_0} \in \mathbf{F};$$

- *furthermore,*

$$\mathscr{P}^n(f_0) \to f_* \text{ in } \mathbf{F};$$

- *there is a structure of an infinite-dimensional complex-analytic manifold on $\mathbf{F}$ which is compatible with the local-uniform norm, in which $\mathscr{P}$ is a contraction.*

We conclude the following.

Corollary 3.6 *The fixed point f_* has an analytic extension to a mapping in $\mathbf{P}_0$ which we will denote in the same way. It is also fixed under $\mathscr{P}$, considered as a transformation $\mathbf{P}_0 \to \mathbf{P}_0$. Moreover, f_* is the unique fixed point of $\mathscr{P}$ in $\mathbf{P}_0$, and for every $f \in \mathbf{P}_0$*

$$\mathscr{P}^n f \to f_*$$

uniformly on compact sets.

Proof By Theorem 3.10, for every $f \in \mathbf{P}_0$ and every $n \in \mathbb{N}$, the renormalization $\mathscr{P}^n(f)$ has a restriction to an analytic map $f_n \in \mathbf{F}$ and $f_n \to f$ in $\mathbf{F}$. By Theorem 3.7, the fixed point $f_* \in \mathbf{P}_0$.

Furthermore, by Lemma 3.9 and Theorem 3.10, for every $f \in \mathbf{P}_0$ the sequence $\mathscr{P}^n(f)$ is precompact in $\mathbf{P}_0$, and hence

$$\mathscr{P}^n(f) \to f_*$$

in $\mathbf{P}_0$ (and hence uniformly on compact subsets of $\mathscr{D}(f_*)$). □

Chapter 4
Numerical Results

Abstract We develop a computational scheme for the parabolic renormalization operator which is based on the asymptotics of the Fatou coordinates at infinity, and apply it to numerical computations of the basin and the domain of the renormalization fixed point and of the spectrum of the parabolic renormalization operator at the fixed point.

Keywords Spectrum of the renormalization operator · Universality

4.1 A Computational Scheme for $\mathscr{P}$

Having mentioned the resurgent properties of the asymptotic expansion of the Fatou coordinate, we proceed to describe the computational scheme for $\mathscr{P}$ (see Fig. 4.1). We begin with a germ of an analytic mapping

$$f(z) = z + z^2 + O(z^3)$$

defined in a neighborhood of the origin. Applying the change of coordinates $w = \kappa(z) = -1/z$, we obtain

$$F(w) = w + 1 + \frac{A}{w} + O\left(\frac{1}{w^2}\right)$$

defined in a neighborhood of ∞. We again use the notation $\Phi_A(w)$ for the function that conjugates F with the unit translation

$$\Phi_A(F(w)) = \Phi_A(w) + 1$$

for $\operatorname{Re} w \gg 1$. We let $\Phi_R(w)$ be the solution of the same functional equation for $\operatorname{Re} w \ll -1$. These changes of coordinate are well-defined up to an additive constant, and

$$\phi_A(z) = \kappa^{-1} \circ \Phi_A \circ \kappa(z), \quad \phi_R(z) = \kappa^{-1} \circ \Phi_R \circ \kappa(z).$$

© The Author(s) 2014
O.E. Lanford III and M. Yampolsky, *Fixed Point of the Parabolic Renormalization Operator*, SpringerBriefs in Mathematics,
DOI 10.1007/978-3-319-11707-2_4

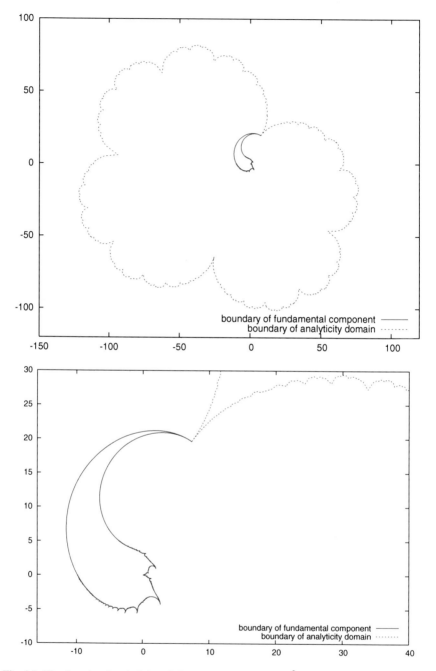

Fig. 4.1 The domain of analyticity of $\mathscr{P} f_0(z)$ for $f_0(z) = z + z^2$, with the immediate parabolic basin indicated

As we have seen in Theorem 2.2, the function $\Phi_A(w)$ has an asymptotic development

$$\Phi_A(w) \sim w - A \log w + \text{const}_A + \sum_{k=1}^{\infty} b_k w^{-k}.$$

The coordinate $\Phi_R(w)$ has an *identical* asymptotic development, differing only by the value of const_R. While this may seem surprising at first glance, recall that these functions are Laplace transforms of *different* analytic continuations of the Borel transform of the same divergent series (plus the $w - A \log w + \text{const}$ term).

We select a large integer M (in practice, $M \approx 100$). We will use the asymptotic expansion to estimate $\Phi_A(w)$ for $w \geq M$ and $\Phi_R(w)$ for $w \leq -M$. Consider an iterate $N \approx 2M$ such that

$$\text{Re } F^N(w) \geq M \quad \text{for Re } w \in [-M-1, -M].$$

Let $v(z)$ be the function

$$v(z) = \text{ixp} \circ \Phi_A \circ F^N \circ (\Phi_R)^{-1} \circ \text{ixp}^{-1}(z).$$

It differs from the parabolic renormalization $\mathscr{P}(f)$ only by rescaling the function and its argument:

$$\mathscr{P}(f)(z) = a_1 v(a_0 z).$$

Now consider a contour Γ connecting $w = -M - 1 + iH$ with $F(w) \approx -M + iH$ which is mapped onto the circle $S_\rho = \{|z| = \rho\}$ for a small value of ρ by $\text{ixp} \circ \Phi_R$. Select $n \in \mathbb{N}$ and consider the n points in S_ρ given by $x_k = \rho \exp(2\pi k/n)$, $k = 0, \ldots, n-1$. We then evaluate the first n coefficients in the Taylor expansion of η at the origin

$$\eta(z) = \sum_{j=0}^{\infty} r_j z^j$$

using a discrete Fourier transform. Specifically, we calculate

$$s_k = v(x_k) \approx \sum_{j=0}^{n-1} r_j (x_k)^j = \sum_{j=0}^{n-1} r_j \rho^j \exp(2\pi k j/n),$$

and apply the inverse discrete Fourier transform:

$$r_j \approx \frac{1}{n\rho^j} \sum_{k=0}^{n-1} s_k \exp(-2\pi k j/n).$$

Since

$$\mathscr{P}(f)(z) = \sum_{j=1}^{\infty} s_j a_1 a_0^j z^j,$$

we have

$$a_1 a_0 s_1 = 1, \text{ and further } a_0 = \frac{s_1}{s_2}.$$

This step completes the computation of the Taylor expansion of $\mathscr{P}(f)$.

4.1.1 Computing f_*

In computing the fixed point $f_*(z)$ we find it more convenient to work with the representation of a germ $f(z) = z + z^2 + \cdots$ in the form

$$f(z) = z \exp(f_{\log}(z)),$$

where $f_{\log}$ is a germ of an analytic function at the origin with $f_{\log}(z) = z + \cdots$. We then rewrite the parabolic renormalization operator in terms of its action on $f_{\log}$:

$$\mathscr{P}_{\log}(f_{\log})(z) = (2\pi i)^{-1} \Phi_A \circ F^N \circ (\Phi_R)^{-1} \circ \mathrm{ixp}^{-1}(z) - \mathrm{ixp}^{-1}(z).$$

This helps to avoid the round-off error which arises from the growth of f_* near the boundary $\partial \mathrm{Dom}(f_*)$.

Modifying the scheme described above for the operator $\mathscr{P}_{\log}$, we calculate the fixed point by iterating $\mathscr{P}$ starting at $f_0(z) = z + z^2$:

Empirical Observation 4.1

$$f_*(z) \approx z + z^2 + 0.(514 - 0.0346i)z^3 + \cdots.$$

Our calculations appear reliable up to the size of the round-off error in double-precision arithmetic ($\sim 10^{-14}$) in the disk of radius $r = 5$ around the origin. As we will see below, the true radius of convergence for the series for f_* is approximately 41 (see the Empirical Observation 4.4).

We also estimated the leading eigenvalue of $D\mathscr{P}|_{f_*}$:

Empirical Observation 4.2 *The eigenvalue of $D\mathscr{P}|_{f_*}$ with the largest modulus is*

$$\lambda \approx -0.017 + 0.040i, \quad |\lambda| \approx 0.044.$$

The small size of λ explains the rapid convergence of the iterates of $\mathscr{P}$ to the fixed point. To obtain this estimate, we write

$$f(z) = z + z^2 + \sum_{k=3}^{\infty} \operatorname{coeff}_k(f)z^k,$$

and consider the spectrum of the $N \times N$ matrix $A = (a_{ij})_{i,j=3\ldots N+3}$, with

$$a_{ij} = \frac{\operatorname{coeff}_j(\mathscr{P}(f_* + \varepsilon z^i)) - \operatorname{coeff}_j(f_*)}{\varepsilon},$$

which serves as a finite-dimensional approximation to $D\mathscr{P}|_{f_*}$.

4.2 Computing the Domain of Analyticity of f_*

4.2.1 Computing the Tail of the Domain $Dom(f_*)$

Computing the tail using an approximate self-similarity near the tip Let us denote

$$t_* \equiv t^{f_*} = \partial \operatorname{Dom}(f_*) \cap \overline{B_0^{f_*}}$$

the endpoint of the tail of the immediate basin of f_*. Let C_R be a repelling fundamental crescent of f_*, and let $w \in C_R$ have the property

$$t_* = \operatorname{ixp} \circ \phi_R(w).$$

Let $k \geq 2$ be such that

$$f_*^k(w) = 0, \text{ so that } f_*^{k-1}(w) = t_*.$$

Denote χ the local branch of $f_*^{-(k-1)}$ which sends t_* to w. Then the composition

$$v \equiv \operatorname{ixp} \circ \phi_R \circ \chi$$

is an analytic map defined in a neighborhood of the endpoint t_*, which fixes it:

$$v(t_*) = t_*.$$

This point can be found numerically:

Empirical Observation 4.3

$$t_* \approx -779.306 - 643.282i, \text{ and } v'(t_*) \approx 0.232 + 0.264i.$$

Thus, we have identified the endpoint of the largest tail of $\text{Dom}(f_*)$. This construction also gives us the means to compute the tail itself. This can be done by successively applying v to the immediate basin $B_0^{f_*}$, thus pulling it in towards t_*.

Now let $q \in C_R$ be any other preimage of 0:

$$f_*^l(q) = t_* \quad \text{for some } q \in \mathbb{N}.$$

Then $v = \text{ixp} \circ \phi_R(q)$ is the endpoint of a different tail in $\partial\text{Dom}(f_*)$. It can be computed by first pulling back the tail of $B_0^{f_*}$ using the inverse branch

$$f_*^{-l} : t_* \mapsto q,$$

and then applying $\text{ixp} \circ \phi_R$.

Computing the tail using the functional equation for an inverse branch A more careful analysis of the tail can be done as follows (Fig. 4.2). Denote by ξ the local branch of f_*^{-1} defined in a slit neighborhood $D_r(0) \setminus [0, r)$ for some small value of r, that sends $0 \mapsto t_*$. We can write the renormalization fixed point equation for this particular branch:

$$\xi = \psi_R \circ \xi \circ \psi_A^{-1}, \tag{4.1}$$

where $\psi_R = \chi \circ \text{ixp} \circ \phi_R$, and ψ_A^{-1} is the appropriately chosen branch of $(\text{ixp} \circ \phi_A)^{-1}$ (thus the "self-similarity" of the tail is exponential, rather than linear). We are going to use the renormalization equation (4.1) *inductively* to compute $\xi(z)$ for sufficiently small values of z, and thus plot the tail.

Representing the numbers in the image of the tail Numerical computations indicate that the value of $r = 0.0002$ is sufficiently small for our needs, and for $|z| < r$ the difference between the left and the right sides of (4.1) is of the order of 10^{-11}. The values of z for which we would like to evaluate $\xi(z)$ become too small to be represented by the standard double precision numbers (and even too small for their logarithms to be so represented). We write

$$s(t) = \exp(2\pi t),$$

and choose $\hat{t}$ so that

$$\exp(-2\pi\hat{t}) = 0.0002, \quad \text{that is } \hat{t} = 1.3555\ldots.$$

We then represent a small positive number x as

$$x = \frac{1}{s^k(t)},$$

for the unique choices of $t \in [\hat{t}, s(\hat{t}))$, and an iterate $k \in \mathbb{N}$.

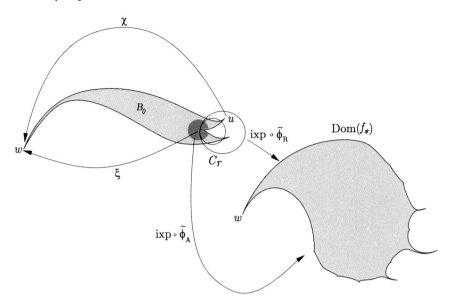

Fig. 4.2 The inverse branches used in computing the tail of Dom(f_*)

We can write any complex number z with $|z| < r$ uniquely as

$$z = (k, t, \theta) \equiv \frac{\exp(2\pi i \theta)}{s^k(t)}, \quad 0 \le \theta < 1.$$

Note that this representation of small numbers makes it very easy to compute logarithms. In particular,

$$\mathrm{ixp}^{-1}((k, t, \theta)) = \theta + i s^{k-1}(t).$$

The next step in applying (4.1) is to apply ϕ_A^{-1} to the right-hand side of the equation. From the first two terms in the asymptotics of

$$\phi_A(z) = -\frac{1}{z} + O(\log |z|) \quad \text{for small } z,$$

it follows that

$$\phi_A^{-1}(y) = -\frac{1}{y + O(\log |y|)} \quad \text{for large } |y|.$$

A numerical estimate shows that for $|y| \ge 10^{18}$, the $O(\log |y|)$ term dissapears into the round-off error when added to y. Thus

$$\psi_A^{-1}((k, t, \theta)) \approx -\frac{1}{\theta + i s^{k-1}(t)} \approx i s^{k-1}(t) = (k - 1, t, 1/4),$$

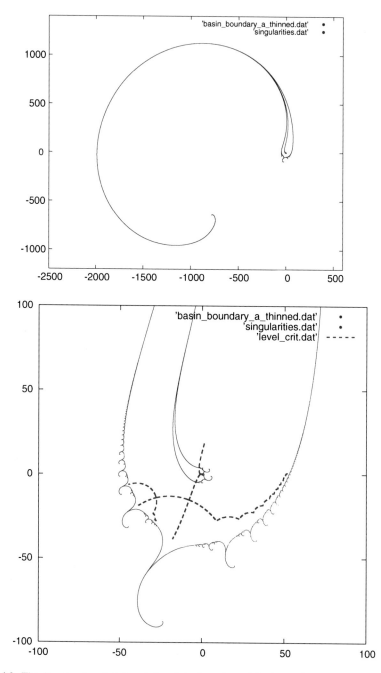

Fig. 4.3 The domain of analyticity of f_* and the boundary of the immediate parabolic basin $B_0^{f_*}$. In the second figure, a part of the critical level curve of f_* is also indicated

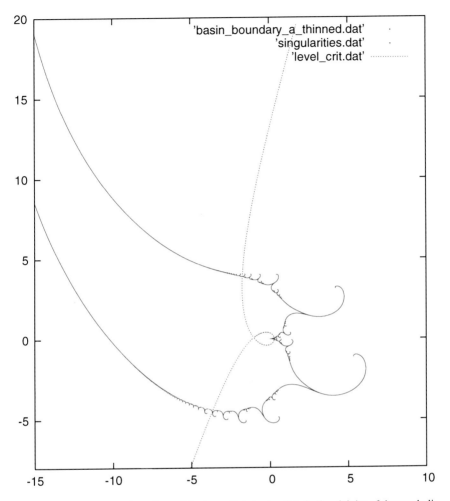

Fig. 4.4 A blow-up of the boundary of the immediate basin of f_* in the vicinity of the parabolic point

provided $s^{k-1}(t) \geq 10^{18}$. A direct estimate shows that for either $k \geq 3$, or $k = 2$ and $t > 18 \log 10/2\pi \approx 6.596$, the last inequality will hold.

The size of the domain of analyticity To draw the pictures of the domain of analyticity of the fixed point of f_* (Figs. 4.3 and 4.4) we employed the following strategy. First, a periodic orbit of period 2 in ∂B_0 was identified. Its preimages give a rough outline of ∂B_0, but become sparse near the "tails", which are not visible in this initial outline. At the next step, the large "tail" of B_0 is computed as described above. Finally, its preimages are used to fill in the remaining gaps in ∂B_0.

As the final step, we calculate the boundary of $\mathrm{Dom}(f_*)$ as

$$\partial \,\mathrm{Dom}(f_*) = \mathrm{ixp} \circ \phi_R(\partial B_0 \cap P_R).$$

An empirical estimate of the inner radius of $\mathrm{Dom}(f_*)$ around the origin allows us to formulate the following observation (see Fig. 4.3):

Empirical Observation 4.4 *The radius of convergence of the Taylor expansion of f_* at the origin is $R \approx 41$.*

Chapter 5
For Dessert: Several Amusing Examples

Abstract We present several borderline examples illustrating the necessity of the conditions in the definition of the renormalization-invariant class.

Keywords Parabolic basin boundary · Mating

5.1 Example of a Map with a Simply-Connected Parabolic Basin Whose Boundary Is Not Locally Connected

Theorem 5.1 *There exists a quadratic rational map* $R : \hat{\mathbb{C}} \to \hat{\mathbb{C}}$ *of degree 2 with the following properties*:

- *R has a simple parabolic fixed point at ∞ with a proper immediate basin B_0^R of degree 2;*
- *the boundary of the immediate basin B_0^R is not locally connected.*

We begin by recalling:

Proposition 5.1 *There exist $\alpha \in \mathbb{R} \setminus \mathbb{Q}$ such that no fixed point of multiplier $e^{2\pi i \alpha}$ for a rational function of degree d can be locally linearizable.*

The first proof of this result is due to Cremer [1927], who gave a sufficient condition for α (see [Mill], Theorem 11.2).

Let us fix α as in Proposition 5.1 and set $\lambda = e^{2\pi i \alpha}$.

Proposition 5.2 *There exists a quadratic rational map R with a simple parabolic point with multiplier 1 at ∞, and a Cremer point with multiplier λ at 0.*

Proof The reader may find a detailed discussion of the dynamics of quadratic rational maps in Milnor's paper [Mil2]. Below we give a brief summary of some relevant facts. Every quadratic rational map F has three fixed points, counted with multiplicity. Let μ_1, μ_2, μ_3 denote the multipliers of the fixed points.

$$\sigma_1 = \mu_1 + \mu_2 + \mu_3, \quad \sigma_2 = \mu_1\mu_2 + \mu_1\mu_3 + \mu_2\mu_3, \quad \sigma_3 = \mu_1\mu_2\mu_3$$

be the elementary symmetric functions of these multipliers. □

© The Author(s) 2014
O.E. Lanford III and M. Yampolsky, *Fixed Point of the Parabolic Renormalization Operator*, SpringerBriefs in Mathematics,
DOI 10.1007/978-3-319-11707-2_5

Proposition 5.3 ([Mil2], Lemma 3.1) *The numbers σ_1, σ_2, σ_3 determine F up to a Möbius conjugacy, and are subject only to the restriction that*

$$\sigma_3 = \sigma_1 - 2.$$

Hence the moduli space of quadratic rational maps up to Möbius conjugacy is canonically isomorphic to $\mathbb{C}^2$, with coordinates σ_1 and σ_2.

Note that for any choice of μ_1, μ_2 with $\mu_1\mu_2 \neq 1$ there exists a quadratic rational map F, unique up to a Möbius conjugacy, which has distinct fixed points with these multipliers. The third multiplier can be computed as $\mu_3 = (2-\mu_1-\mu_2)/(1-\mu_1\mu_2)$.

Thus there exists a quadratic rational map R with fixed points at 0, ∞ such that $R'(0) = \lambda$ and $R'(\infty) = 1$. The map R has only two critical orbits, and at least one of them has to contain the Cremer fixed point 0 in its closure. Thus the parabolic basin of ∞ can contain only one critical value, and hence ∞ is a simple parabolic point.

Let us fix R as in Proposition 5.2. The following is an immediate consequence of Montel's Theorem:

Proposition 5.4 *The Julia set*
$$J(R) = \partial B_0^R.$$

Let us now argue by way of contradiction and assume that ∂B_0^R is locally connected. It is elementary to see that B_0^R is simply-connected. Denote $D : \mathbb{T} \to \mathbb{T}$ the doubling map $D(x) = 2x \mod 1$.

Proposition 5.5 *There exists a continuous surjective map $\gamma : \mathbb{T} \to \partial B_0^R$ such that*

$$\gamma \circ D = R \circ \gamma.$$

Proof We use the fact that R is conformally conjugate to the Koebe function K on the immediate basin B_0^R,

$$\psi \circ R \circ \psi^{-1} = K,$$

and apply the Carathéodory Theorem to the conformal map ψ.   □

Let us denote $W = \gamma^{-1}(0)$. We claim:

Proposition 5.6 *The set of angles W is finite.*

To prove this first note that R is a local homeomorphism at 0 and therefore it induces a homeomorphism on W. The proposition now follows from the following general fact (see e.g. [Mill], Lemma 18.8):

Lemma 5.1 *Let X be a compact metric space and let $h : X \to X$ be a homeomorphism. If h is expanding then X is finite.*

We conclude:

Corollary 5.1 *There exists a periodic angle $x \in W$.*

We are now set to prove Theorem 5.1:

Proof Let ψ be as in Proposition 5.5 and let

$$\ell = \psi^{-1}(\mathbb{R}) \subset B_0^R.$$

Let us denote U^+, U^- the components of $B_0^R \setminus \ell$. To fix the ideas, assume that the angle x from Proposition 5.1 corresponds to a prime end p in the Carathéodory completion of U^+, and has period m under the doubling map. Let $Q : U^+ \to U^+$ be the branch of R^{-m} whose Carathéodory extension fixes p. By the Denjoy-Wolff Theorem, all orbits of Q converge to 0. Hence, 0 has a nonempty basin under Q, and thus is either repelling or parabolic for R, which contradicts our assumptions.   □

We finally note, that up to a conjugacy by $z \mapsto az$ with $a \in \mathbb{C}^*$, our map R has the form

$$R(z) = \frac{z^2 + \lambda z}{z + 1}.$$

5.2 Example of a Map in $\mathbf{P} \setminus \mathbf{P_0}$

We begin by showing:

Theorem 5.2 *There exists a quadratic rational map R with a simple parabolic fixed point whose immediate basin B_0^R is simply connected and has a locally connected boundary, which is not a Jordan curve.*

We find our example in the family of quadratic rational maps

$$R_a(z) = \frac{a}{z^2 + 2z},$$

which was considered in [AY]. For $a \neq 0$ the map R_a has two simple critical points $c_1 = -1$ and $c_2 = \infty$. The latter one is periodic with period 2:

$$\infty \mapsto 0 \mapsto \infty.$$

Fix $a_0 = 32/27$ and let $R \equiv R_{a_0}$. The rational function R has a parabolic fixed point $z_0 = -4/3$ with multiplier 1. Since there is a single critical orbit of R which may be attracted by z_0, the point z_0 is a simple parabolic fixed point. Elementary considerations imply that B_0^R is simply-connected, it is bounded since a neighborhood of ∞ belongs to the basin of the super-attracting cycle of period 2.

By Montel's Theorem $\partial B_0^R = J(R)$. The rational map R can be described as a conformal mating of the basilica $z \mapsto z^2 - 1$ and the parabolic map $z \mapsto z + z^2$. The Julia set of R is locally connected by [TY].

The other fixed point of R is $\alpha = 2/3$. It is repelling: $R'(\alpha) = -5/4$. It is elementary to verify that the interval $[0, \alpha]$ is invariant under the second iterate of R, and that

$$R^2(x) < x \text{ for } x \in (0, \alpha).$$

Hence, the interval $(0, \alpha)$ belongs to the super-attracting basin, and so does the interval (α, ∞). The point α is on the boundary of the immediate basin B_0^R and hence must be accessible from B_0^R. By real symmetry, α is bi-accessible from B_0^R. Consider the parabolic renormalization $F \equiv \mathscr{P}(R)$. Preimages of α are dense in ∂B_0^R, and therefore the boundary $\mathscr{D}(F)$ contains locally conformal preimages of a neighborhood of α in $J(R)$. In particular, $\partial \mathscr{D}(F)$ contains bi-accessible points, and therefore is not Jordan. On the other hand, $\partial \mathscr{D}(F)$ is a locally homeomorphic image of a locally connected set, and hence is locally connected. Thus, $F \in \mathbf{P} \setminus \mathbf{P}_0$. Note that by Theorem 3.9, the parabolic renormalization $\mathscr{P}(F) \in \mathbf{P}_0$.

References

[AY]   M. Aspenberg, M. Yampolsky, Mating non-renormalizable quadratic polynomials. Commun. Math. Phys. **287**, 1–40 (2009)

[BE]   X. Buff, A. Epstein, A parabolic Pommerenke-Levin-Yoccoz inequality. Fund. Math. **172**(3), 249–289 (2002)

[BH]   X. Buff, J.H. Hubbard, *Dynamics in One Complex Variable*. Manuscript

[BS]   X. Buff, A. Chéritat, Quadratic Julia sets with positive area. Ann. Math. **176**, 673–746 (2012)

[Co]   J.B. Conway, *Functions of One Complex Variable II*. Graduate Texts in Mathematics, vol. 159 (Springer, New York, 1995)

[Do1]   A. Douady, Systèmes dynamiques holomorphes. Séminaire Bourbaki, Astérisque **105–106**, 39–64 (1983)

[Do2]   A. Douady, Does a Julia set depend continuously on the polynomial? In Complex dynamical systems: the mathematics behind the Mandelbrot set and Julia sets. in *Proceedings of Symposia in Applied Mathematics*, ed. by R.L. Devaney, vol. 49 (American Mathematical Society, Providence, 1994), pp. 91–138

[DH]   A. Douady, J.H. Hubbard, Etude dynamique des polynômes complexes, I-II. Pub. Math. d'Orsay (1984)

[Du]   A. Dudko, Dynamics of holomorphic maps: resurgence of Fatou coordinates, and poly-time computability of Julia sets. PhD Thesis (University of Toronto, 2012)

[DS1]   A. Dudko, D. Sauzin, The resurgent character of the Fatou coordinates of a simple parabolic germ. C. R. Math. **352**, 255–261 (2014)

[DS2]   A. Dudko, D. Sauzin, On the resurgent approach to Écalle-Voronin's invariants. Preprint arXiv:1307.8095

[Ec]   J. Écalle, Les fonctions résurgentes, Publ. Math. d'Orsay [vol. 1: 81–05, vol. 2: 81–06, vol. 3: 85–05] 1981, 1985

[Ep]   A. Epstein, Towers of finite type complex analytic maps, PhD Thesis (CUNY, 1993)

[EKT]   A. Epstein, L. Keen, C. Tresser, The set of maps $F_{a,b} : x \mapsto x + a + \frac{b}{2\pi} \sin(2\pi x)$ with any given rotation interval is contractible. Commun. Math. Phys. **173**, 313–333 (1995)

[EY]   A. Epstein, M. Yampolsky, The universal parabolic map. Accepted to Ergodic Th. Dynam. Sys. http://www.math.toronto.edu/yampol

[IS]   H. Inou, M. Shishikura, The renormalization for parabolic fixed points and their perturbation. Preprint (2008)

[Lan]   O.E. Lanford, *The zero rotation number fixed point*. Manuscript

[Lav]   P. Lavaurs, *Systèmes dynamiques holomorphes: Explosion de points périodiques* (Université de Paris-Sud, Thèse, 1989)

© The Author(s) 2014

O.E. Lanford III and M. Yampolsky, *Fixed Point of the Parabolic Renormalization Operator*, SpringerBriefs in Mathematics, DOI 10.1007/978-3-319-11707-2

[Lyu]   M. Lyubich, The dynamics of rational transforms: the topological picture. Russ. Math. Surv. **41**(4), 43–117 (1986)

[Mil1] J. Milnor, *Dynamics in One Complex Variable*, 3rd edn. Annals of Mathematics Studies, vol. 160 (Princeton University Press, Princeton, 2006)

[Mil2] J. Milnor, Geometry and dynamics of quadratic rational maps. Exp. Math. **2**, 37–83 (1993)

[Mil3] J. Milnor, Pasting together Julia sets—a worked out example of mating. Exp. Math. **13**(1), 55–92 (2004)

[Ram]   J.-P. Ramis, *Séries divergentes et théories asymptotiques*, Panoramas et Synthèses (1994)

[Sau]   D. Sauzin, *Resurgent Functions and Splitting Problems*, vol. 1493 (RIMS Kokyuroku, 2006), pp. 48–117, http://arxiv.org/abs/0706.0137

[Sh]   M. Shishikura, The Hausdorff dimension of the boundary of the Mandelbrot set and Julia sets. Ann. Math. **147**(2), 225–267 (1998)

[TY]   L. Tan, Y. Yin, Local connectivity of the Julia set for geometrically finite rational maps. Sci. China (Serie A) **39**, 39–47 (1996)

[Vor]   S.M. Voronin, Analytic classification of germs of conformal maps $(\mathbb{C}, 0) \to (\mathbb{C}, 0)$ with identity linear part. Funct. Anal. Appl. **15**, 1–17 (1981)

[Ya1]   M. Yampolsky, The attractor of renormalization and rigidity of towers of critical circle maps. Commun. Math. Phys. **218**(3), 537–568 (2001)

[YZ]   M. Yampolsky, S. Zakeri, Mating Siegel quadratic polynomials. J. Am. Math. Soc. **14**, 25–78 (2000)

Index

© The Author(s) 2014
O.E. Lanford III and M. Yampolsky, *Fixed Point of the Parabolic Renormalization Operator*, SpringerBriefs in Mathematics,
DOI 10.1007/978-3-319-11707-2